COLLINS & BROWN

ULTIMATE

WOOD

WORK

BIBLE

A Complete Reference With Step-By-Step Techniques

Phil Davy and Ben Plewes

First published in the United Kingdom in 2017 by
Collins & Brown
43 Great Ormond Street
London WC1N 3HZ

An imprint of Pavilion Books Company Ltd

Distributed in the United States and Canada by
Sterling Publishing Co
1166 Avenue of the Americas
New York, NY 10036

ISBN 978-1-911163-43-5

A CIP catalogue for this book is available from the British library.

10 9 8 7 6 5 4 3 2 1

Reproduction by Mission Productions, Hong Kong
Printed and bound by COS Printing, Singapore

This book can be ordered direct from the publisher at
www.pavilionbooks.com

ULT

WORK

BIBLE

contents

introduction

Wood is extraordinary – in fact, there are few natural materials to rival it. What other resource provides food and shelter for wildlife, fuel for heating and material for house construction, but can equally be used for making fine furniture, boats and musical instruments, or even for firing the summer barbecue?

Wood is full of surprises, too. That dusty old board at the sawmill, which seems little more than firewood quality, may suddenly be transformed when you run a plane across its surface. Suddenly it becomes full of life and colour – you don't get that excitement from too many other materials.

No two pieces of timber (lumber) are exactly the same, even when cut from the same board, which can often provide a challenge for woodworkers. For example, the grain at one end of a piece of softwood may behave very differently from that at the other end, which can be frustrating when you are a beginner. But practise using plane and chisel and these same differences can open up a wealth of opportunities; all part of the learning curve when you are just starting out. And no matter how many years ago you may have first picked up a woodworking tool, there is always something new to learn, whether it is experimenting with a finish, discovering the quirks of a traditional hand tool, or getting to grips with a new workshop machine.

Discovering hardwoods is like opening a door to another world. Each wood has its own characteristics and aesthetic qualities. Look at one side of a piece of quartersawn English oak, say, and you may find the open grain warm but rather uniform. Check the adjacent face and you are likely to see the most beautiful flecks, or rays, across the surface.

Then of course there is the welcoming aroma when you first enter your workshop. Even a lowly piece of pine can smell delicious when you have taken a few fresh shavings with a plane. Mixed with the heady brew of traditional finishing products, there is nowhere quite like the atmosphere of a woodwork shop.

If you are discovering woodworking for the first time, welcome to the family! This book aims to guide you through the process of choosing tools and planning a workshop – which may start as a portable workbench in the kitchen and perhaps later become that fully equipped custom shop we may dream about. However, you cannot beat hands-on tuition so if you are really a beginner, then check out the range of local evening classes or short courses for woodworkers in your area. In this digital age there are also plenty of substitutes if time is limited, such as DVDs and online forums. Woodworkers tend to be a friendly bunch, so you can bet that there will be someone out there only too happy to give advice or guidance. But be warned – woodworking is addictive! Once you have started, you may never want to stop …

Phil Davy

Ben Plewes

the essential workshop

For many woodworkers the workshop offers a place to escape the pressures of the outside world, but for it to be an effective work environment as well requires careful planning, especially if space and budget are tight. If starting out with an empty workshop it makes sense to get the layout right from the outset. Making an existing workshop safer and more efficient may be more of a challenge but it is worthwhile, whether woodwork is your hobby or perhaps the start of a new career.

location and planning

Most of us do not have the budget for a dream workshop, but a garage, shed, attic, basement or even a spare room are all suitable locations for a home workshop. Think about the type of woodwork you want to carry out initially, because workshop size is an essential consideration. A model maker or a musical instrument maker – using mostly hand tools – may only need space for a workbench plus one or two compact machines, so a spare room or basement could be perfect. If you intend to plane up your own wood from rough-sawn boards, however, then you will need a bigger workspace for machinery, including an area for storing timber (lumber).

Planning permission

Before you begin ordering any building materials, or invest in a profabricated shed, check with your local authority whether you need any permissions or approvals. A wooden shed may not need permission if it complies with height restrictions and has an area relative to the overall size of the garden (yard), but using any workshop for commercial purposes in a residential area will definitely need permission, so it is essential to find out where you stand first.

Related info

Layout and workflow (see page 12)
Security and power (see page 14)
Safer woodwork (see page 22)
Workshop storage (see page 18)

Wooden shed

You can turn a simple wooden shed into a comfortable small workspace quite easily, especially if budget is limited. The basic wooden shed is a relatively cheap option, though its lifespan will be shorter than a garage or brick building and it will need more maintenance. Once a suitable base is laid, you can erect the structure and make it weatherproof in a day. Staple a plastic membrane around the inside walls to restrict the passage of damp and condensation, then add a layer of insulation. Finally, lining walls with MDF or plywood will increase overall rigidity and create a firm backing, so you can attach shelves and storage cupboards later.

Unless you build one from scratch, most prefabricated sheds are quite insubstantial. Normally constructed from shiplap or tongue-and-groove boards, they are fairly easy to adapt or repair. The downside is that security can be a problem, but you can replace doors and windows with sturdier joinery at a later date. It is important to fit plastic guttering to deal with rainwater run-off.

In a large garden (yard) you may consider siting the workshop away from the house, but this will not seem a good idea when bad weather strikes. For power supply and security issues it also makes sense to locate a workshop as near a house as possible, although a wooden building should be no closer than 2m (6ft).

Garage

A garage or outbuilding can make an excellent small workshop, convenient for unloading timber (lumber) and materials outside. It is usually close to the house and brick or blockwork construction makes it fairly secure, while there is likely to be a power supply already. Relatively easy to insulate, a garage usually has a concrete floor, perfect for machinery but tiring on the feet after hours at the bench. Mount machines on castors so you can move them around for each operation. Remember to insulate the door as well, whether it is the up-and-over type or a more traditional hinged version.

If the workspace also needs to house a car you may need to build a bench that folds away after use, while storage will require clever planning. There may be sufficient height to suspend timber (lumber) from the ceiling, while tools could be wall-mounted or kept in cupboards. A lack of windows improves security but means that overhead lighting is essential – to benefit from daylight, consider fitting a skylight to the roof.

Custom workshop

Building your ultimate workshop is quite feasible if budget and available space allow. Although undoubtedly costly, this means you can design a workspace to accommodate any future woodworking needs, with almost endless door and window options. You can choose materials and construction methods – although you may find a substantial timber (lumber) structure easier to build yourself and it is possible that your local planning authority may not allow some materials to be used. Estimating cost and quantities is critical if working to a budget, and for a complex building it may be worth using an architect to prepare plans. Once you have decided on the maximum size of the workshop, play around with sketches on paper or computer.

Renting

If what is currently a leisure activity becomes a business, you may need to rent a workshop. You may just need the extra space, but if you are self-employed and will accept payment for your work ordinary household insurance will rarely cover a workshop operating as a business. To reduce your overhead costs as you build up your business you may be able to share a workspace with other craft workers.

Spare room

Probably the cheapest option for setting up a workshop is to use a spare room in the house, where security is likely to be much better than for any structure in the garden. For some woodwork, such as model making or luthiery, you may just need space for a bench and very little else in the way of storage or machinery. Running costs will be much lower because the heating will be shared with the rest of the house. A window means you will have natural light. though you could also consider adding secondary glazing if necessary.

layout and workflow

To make efficient use of the workshop it makes sense to plan the workflow, especially if machinery is involved. Otherwise you might find that a workbench sited in the wrong position, for instance, obstructs a length of timber (lumber) exiting a table saw or thicknesser (thickness planer). After having invested considerable time and money on installation, you don't want to find out later that you need to reposition equipment or perhaps rethink the whole way you are working.

Related info

Location and planning (see page 10)
Workshop storage (see page 18)
Workbench basics (see page 16)
Safer woodwork (see page 22)

Computer or paper plans

With a basic drawing program you can experiment with workshop layout on a computer. This could just be a basic floor plan or to show more complex 3D views, depending on your software. Once you have mastered the basics, altering the layout should be simple.

If you do not have the right software, planning is very easy to do on paper instead. Measure up the area and then make a scale drawing of the floor plan, including the position of windows and doors. Using a scale of 1:20 (or 1 inch to 2 feet in imperial), you should be able to fit an average-size workshop on a sheet of A4 (letter) paper. Graph paper makes drawing the plan easier and this can be downloaded from the Internet in either metric or imperial formats. Using a calculator to scale down sizes as necessary, draw workbench and machinery at the same scale on coloured paper and cut out. Move these coloured cutouts around the plan until you arrive at the best layout.

Workflow

If there is sufficient space within the area, machines such as a table saw or planer thicknesser (jointer planer) should ideally be positioned in the centre. To feed timber (lumber) through these machines you will need the same distance behind as in front. In a compact shop, place a planer (jointer against a wall or make use of an open doorway, though this will increase noise levels outside. A mitre saw should ideally be central along the wall, although if you will be mainly crosscutting boards, make sure there is more space available to the left side. For cutting sheet materials, you will need space all round a table saw, although you may have to cut up full-size boards outside the workshop if the space is too cramped.

Consider siting equipment diagonally to make best use of space. In a small workshop the bench will need to be against a wall, preferably by a window. If there is plenty of space, however, a bench in the middle of the floor would mean that you have access all the way round. Think hard about what you will be making: a few machined components may not take up much space, but the resulting glued-up dining table will require a much larger area.

Another thing to consider is the working height of machinery. The table on a bandsaw is higher than on other machines, so if space is tight timber (lumber) can be fed above an adjacent saw or planer (jointer). When machining timber (lumber) in a sequence, check that different table heights on separate machines will not lead to fouling as a board travels through the system. If necessary, raise lower machines on blocks of wood so tables are at a uniform height, although you may only need to tilt a machine slightly to solve a workflow problem. When ripping a long board, you may be able to pass it out through a convenient window. If possible, bolt small machines to mobile bases and you will find a way to cope with that extra long board.

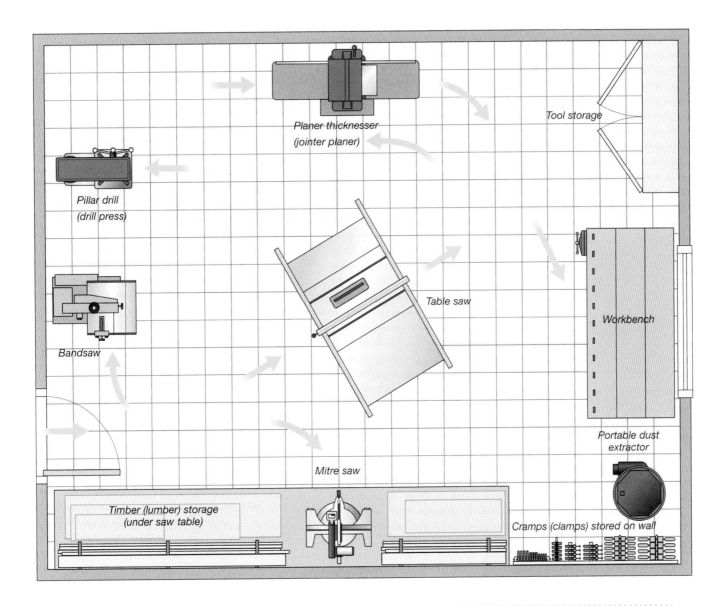

Planer thicknesser
(jointer planer)

Tool storage

Pillar drill
(drill press)

Table saw

Workbench

Bandsaw

Portable dust
extractor

Mitre saw

Timber (lumber) storage
(under saw table)

Cramps (clamps) stored on wall

The ideal workshop

Make space in a corner of the workspace for a sharpening area, to include a bench grinder and stones for honing hand tools. Keep this station away from your main workbench so there is no chance of oil or water coming into contact with your work in progress. Cover stones when not in use to keep sawdust at bay.

A table saw or combination machine installed in a small workshop can have a piece of MDF or plywood fitted over the table to create an extra assembly surface. The blade will need to be lowered and guards removed, so make sure these are replaced correctly before using the machine.

Keep a torch handy close to the door in case of power cuts.

security and power

Once the workshop is finished, making it secure should be the next priority – there are plenty of steps you can take to deter intruders and protect the contents. Adding power and lighting requires careful planning and should only be carried out by a qualified electrician. With increasing energy costs, efficient heating is also important; a cold, damp workshop is not conducive to comfortable working or to safe storage of tools and materials.

Safe and secure

It is easy to ignore security and assume a break-in will never happen, but a workshop filled with expensive hand or power tools is a potential target. You need to make it as difficult as possible for a potential thief to enter; to assess the risks, work out how you would get inside if you were locked out without a key. Could an intruder lever off the door hinges, climb in through a window or saw through that cheap padlock? A workshop sited away from the house can be more at risk than one nearby, though both are vulnerable. Thefts are often opportunist break-ins, so any deterrent is worthwhile – especially if it creates noise to scare off an intruder.

Workshop alarms – these are inexpensive and easy to fit and include motion sensor devices, contact alarms for doors and windows, and PIR systems, which switch on a light when movement is detected. The more sophisticated PIR devices record short bursts of video and static images.

Window security – replacing glass in windows with clear acrylic or polycarbonate sheeting can be quite expensive, but sheets

> ### Related info
> Location and planning (see page 10)
> Layout and workflow (see page 12)
> Workshop storage (see page 18)
> Workbench basics (see page 16)
> Safer woodwork (see page 22)
> Acclimatizing timber (lumber) (see page 162)

cannot be broken. Alternatively make a set of internal shutters from plywood or MDF, which can be lifted into position at the end of a session, or fit steel bars or grilles – although for several windows the cost could be high.

Identify tools – power tools are particularly attractive to thieves because they are very easy to dispose of quickly. To make them easier to identify, use a mini tool to engrave them discreetly with your postcode (zip code), or use a security dot marking system that enables you to register equipment on the Internet.

Contact alarm fitted inside door.

Passive infrared security light.

Padlock with motion sensor alarm.

Insurance

Many companies will include insurance cover for a shed or outbuilding on a normal household policy, but it only takes a few power tools or machines to reach the maximum limit for contents. Check the value of your equipment and see if you can increase the contents limit if necessary. If you start to make money from your woodworking skills, you will probably find that you must have business insurance in place.

Power and light

Unless your workshop is directly attached to the main house, you will need to install a power supply. The safest option is to use armoured cable buried underground, but contact your local government office to check current rules and regulations. An easier method of getting the power across an intervening space is to suspend the cable from an overhead catenary wire, which should hang at a minimum height to reduce the risk of accidental damage and must be correctly earthed. Consult a qualified electrician for the fitting and inspecting of any electrical work, always use waterproof connectors for exterior installations and make sure all work complies with current regulations.

Workshop lighting should be adaptable to suit different working conditions. Low energy bulbs are fine as general lighting, although you may also want to fit a couple of fluorescent strip lights overhead in case you need additional light. For benchwork, a srnall office desk light can be handy, and there are clip-on versions that are feasible for illuminating the working area of machines. Low wattage LEDs provide bright light and are cheap to run. To make best use of any available daylight, place your main workbench close to the window.

You will probably need more electric sockets than you anticipate; site them along walls at regular intervals so that power tool cables do not run across the floor creating a hazard.

Insulation and heating

Insulating an outbuilding or garage will not only reduce noise but also save money on heating bills, which is an important factor with today's increasing energy costs. You should insulate floor, ceiling and walls; the cheapest method is to use the recycled plastic or fibreglass rolls that are sold as house roofing insulation – avoid polystyrene because this does not really reduce noise levels effectively.

Unless you are likely to use a workshop only during summer months, some form of heating will be necessary. A low-output electric background heater is designed to be left switched on and can be used with a timer device to reduce running costs. Consider using a stove fuelled by wood waste if you are using the workshop for any length of time.

You cannot hope to carry out fine woodwork in a damp workshop so you may need to add a damp-proof membrane to limit condensation. Humidity levels should be controlled to stabilize the moisture content of timber (lumber) and prevent tools deteriorating – a small dehumidifier will reduce moisture and keep the air warm. Any heater that burns liquid fuel gives off moisture, so think carefully before using a portable gas heater – which may be convenient, but is also expensive to run.

Making the most of the light

You can reflect much more natural light into your workshop area by painting all the internal walls and the ceiling white.

workbench basics

Whether you build it yourself or just buy one, a decent workbench should be sturdy enough to withstand the heaviest mallet blows and support the most substantial lengths of timber (lumber) adequately. Building a workbench is an ideal project for a novice woodworker to learn timber (lumber) preparation, basic jointing and assembly techniques. A reliable workbench does not need to be complex, although making your own means it can be designed to suit your specific needs.

Bench basics

A solid bench is essentially a large workstation that enables you to prepare timber (lumber) or work on a small project without any vibration or movement, whether the workpiece is held vertically, horizontally or tilted in the vice. It is constructed with suitable framing joints, and there is normally space underneath for storing timber (lumber) or work-in-progress. You can enclose this space and fit doors at ends or front to make it into a secure cupboard for tools, which helps to add weight and stability to the structure. A drawer below the top is ideal to keep marking tools safe and close at hand.

Any close-grained hardwood is suitable for making a bench; beech is traditionally the favourite choice, although it is not as stable a wood as maple. If budget is very tight, build a softwood framework, with legs 100 x 100mm (4 x 4in) PAR in section. For weight and robustness make the top at least 50mm (2in) thick, ideally from hardwood. The surface should be absolutely flat, because you need to use it when planing timber (lumber) flat and true. While the top surface of a cabinetmaker's bench is usually completely level, a carpenter's version generally has a recessed tool well running along the back. This means you can lay tools down temporarily without them fouling work on the bench top itself. For convenient access to narrow hand tools – such as chisels, squares and saws – you can fix a vertical storage rack along the rear edge. Where space is really limited, consider building a bench that you can fold flat after work is finished – this could be the ideal answer in a garage workshop that is shared with a car.

Bench ergonomics

As well as being solid and rigid, the top of the bench must be at the correct working height otherwise you may develop back problems or simply not be able to control tools efficiently. Use wooden blocks to increase the height of a low bench, or cut down the legs if it is too high for comfortable working. If possble, try out other benches first to confirm a satisfactory height; for a person of about average height it should be 860–915mm (34–36in).

Related info
Location and planning
(see page 10)
Layout and workflow
(see page 12)
Workshop storage
(see page 18)
Construction methods
(see pages 178–213)

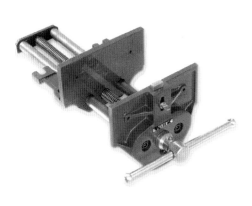

instead of sash cramps (clamps) for gluing components together.

Vices are either cast iron (English pattern) and bolted beneath the bench top, or made from hardwood on continental-style benches. It is worth paying a bit more for a quick-release vice, because jaw opening or closing is much faster than on the standard version with a screw thread mechanism. To prevent damage to tools or timber (lumber), fit hardwood facings to inside jaw faces.

Bench vices

You should have at least one vice on the bench for gripping timber (lumber) while planing. If you are building your own bench you can position a vice wherever you choose, but if you are right-handed the front vice is normally bolted on the left, or on the right if you are left-handed. To increase the cramping (clamping) capacity on a traditional cabinetmaker's bench, sometimes a tail vice can be fitted at the far end. Two rows of square or round holes along the top of the bench will enable you to use this vice with matching bench dogs. These may be hardwood, steel or plastic and are simply inserted where convenient to act as stops. The workpiece is placed between the dogs, and the vice adjusted to hold it securely for tasks such as planing, routing or sanding. Bench dogs can also be used

Folding workbenches

Many woodworkers begin with a portable DIY workbench such as a Workmate and these make convenient temporary work surfaces for tasks such as assembling carcasses, mounting small machines and working outdoors. They are designed to be folded and stowed in a car boot (trunk), and are relied on by tradesmen worldwide. Another option is a pair of sawhorses, which are cheap to make, sturdy and can be stacked.

Holding devices

A bench stop is useful if you have no bench dog holes; it can be raised above the bench top as a stop when planing timber (lumber) and lowered again when not needed. For gripping work on the bench top, a holdfast is a versatile cramping device, particularly for woodcarvers; it is often supplied with a steel collar inset into the surface. Easy to make but incredibly useful, a bench hook prevents damage to the bench when using a back saw. It can be tightened in the vice or held against the bench top. A shooting board is another bench-mounted device that can be either bought or made in the workshop from hardwood.

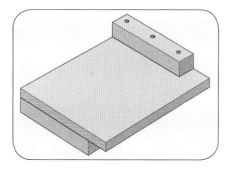

The illustration shows a bench hook for right-handed use – for left-handed use simply reverse the gap at the end of each batten. It should measure approximately 250 x 150mm (10 x 6in) and be 25mm (1in) thick, with a 30mm (1¼in) square section batten fixed at opposite ends – one on the top and one on the bottom.

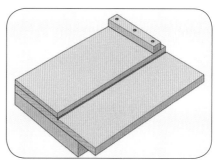

Shooting boards consist of two main pieces secured together, with the top section narrower than the bottom to form a sliding platform for a bench plane to run on its side. A solid wood batten secured at right angles holds workpieces in place. Approximate measurements for a good shooting board are 610 x 230mm (24 x 9in) overall with 75mm (3in) of the width forming the bench plane platform.

workshop storage

A tidy, organized workshop means greater working efficiency. Storage for tools and materials is a priority so you do not waste time searching for mislaid items – remember, if you cannot find it you won't use it! Even if you are naturally untidy, try to get into the habit of tidying up regularly. If you have to sweep the workbench before you start each day it can soon become demoralizing.

Materials

Unless you have a huge workshop where you can store sheet materials stacked flat, they must be stored vertically on edge in a rack. This will give easy access and enables you to pull out a sheet as needed. To prevent the boards warping, wedge them together with wooden blocks. Spacers between boards will also allow air to circulate and keep them stable. Softwood lengths should preferably be stored horizontally because they will soon bow if stood on end. If possible, stack long, heavy boards flat on the workshop floor, or if space is tight they can be stacked on substantial shelving. Apart from heavy timber (lumber), always try to keep the floor area as clear from obstructions as possible.

Sturdy steel shelving can be positioned along one side of the workshop – it can be quite narrow if necessary. Adjustable brackets mean that you can alter the shelf spacing easily to suit future needs. A cheaper alternative is to make your own gallows brackets from softwood. Cut a sheet of MDF or shuttering plywood to width for the actual shelves, which should always be screwed to the brackets for safety. This makes a good recycling project, because boards do not have to be new, although they should be flat. Smaller pieces of timber (lumber), power tools and projects can also be stored on these shelves.

Second-hand kitchen units fitted around the wall provide rigidity and can be used to support smaller machines such as bench grinders and bench drills, as well as providing storage for power tools, abrasives and tooling. Reclaimed hollow interior doors make ideal auxiliary worktops, although they are not substantial enough to act as a main bench – thick MDF is a stronger substitute for bench tops. Lightweight doors ripped down the middle are also suitable for shelving, although you will probably need to reinforce the sawn edges with strips of batten.

Most woodworkers are hoarders, retaining tiny pieces of wood for that proverbial rainy day, which may never come! If you are working with expensive hardwoods you will probably want to keep all but the smallest off-cuts; stack them neatly by length or species, and throw unusable off-cuts in a plastic bin. Make sure the bin is cleared out regularly – if you don't have a wood burner yourself, give this waste material to someone who can use it as fuel.

Hardware and finishes

Wood finishes are a fire risk, so store tins of stains and varnish in a locked steel cabinet, if possible, and keep a fire extinguisher close to the workshop door. Stack boxes of screws neatly, while pins and fasteners can be housed in plastic drawer systems fixed to the wall. Alternatively, use recycled glass preserve jars, labelled according to size and gauge of the contents. Make a simple open box with a handle for carrying nails around.

Storing tools

Hand tools can be stored in cupboards, drawers or toolboxes. Many woodworkers build beautiful cabinets to accommodate their entire tool collections – if you own some fine hand tools, this could be a great poject to keep them secure.

Often-used tools should be stored close to the bench if possible, with chisels either in a bench rack, in a drawer or on the wall. A simple shadow board will enable you to see exactly where a tool belongs – and to see immediately if a tool is missing. To keep rust at bay, keep silica gel packs in drawers or wrap heirloom tools in rust-inhibiting paper.

Professional power tools are usually supplied with a dedicated storage case. If they are only used occasionally, store them under the bench or in a cupboard. For more frequent use, devote a section of shelving to them or an entire cupboard. Making a row of pigeonholes will keep cables tidy. Drill a block of wood with a series of holes for inserting router cutters, so you can see the one you need at a glance.

Related info
Location and planning (see page 10)
Safer woodwork (see page 22)
Caring for stored wood (see page 133)
Glue storage (see page 220)

French cleats

A cupboard full of tools will be incredibly heavy, so you need to be sure that it is fitted securely to the workshop wall. The strongest fixing method is to use French cleats, a clever system that is very simple to make: it consists of two wooden battens butting against each other at 45 degrees. Screw the lower cleat to the wall and the upper one to the back of the cupboard, and then just lift the cupboard onto the wall cleat. If you make several cupboards the same size, you can move them to different locations around the workshop in future, if required.

Cramps (clamps)

It makes sense to group cramps (clamps) together, because you may sometimes need to access them all when gluing up a large project. If your shop has a large number, make a trolley to house them so that they can be moved closer to the job when needed. In a small workshop, a wall may be the only storage space available but this is ideal for storing long sash cramps (clamps) neatly suspended from dowels, pegs or a dedicated rack, as shown below. After use, always return cramps (clamps) to their storage place.

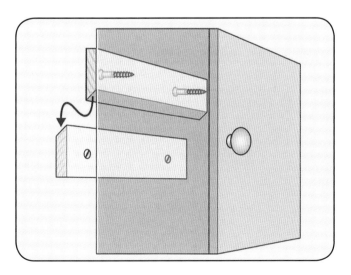

dust control

It is impossible to work with wood or sheet materials and produce no waste, whether this is off-cuts, chippings or sawdust. Controlling this waste efficiently should be a priority in any workshop – not only can dust adversely affect your health, but it creates a fire hazard if not dealt with effectively at source.

Controlling dust

Whether you are working predominantly with hand tools or machines, efficient dust control is important. Sanding by hand may not produce as much dust as an electric sander, but you should still take precautions. Wear a face mask when working with MDF or certain hardwoods. At the end of a working session, clear up sawdust and empty it into plastic sacks for disposal.

Power tools are notorious for producing plenty of mess, particularly routers, planers (jointers) and sanders. Some tools are equipped with dust bags, although routers produce copious amounts of dust and should ideally be fitted with a flexible hose attached to a dust extractor. Machines such as planer thicknessers (jointer planers) will eject large amounts of waste and without extraction will clog up quickly, which leads to poor timber (lumber) feed rate, strain on motors and a less-than-perfect surface finish.

> **Related info**
> Location and planning (see page 10)
> Layout and workflow (see page 12)
> Safer woodwork (see page 22)
> Power tools (see pages 60–89)
> Woodworking machines (see pages 90–119)

Vacuum extractors

A small portable vacuum extractor is handy for clearing up mess at the end of a session and can be used generally to keep the shop clean. A dedicated workshop unit will have a flexible hose that you connect to the outlet of a power tool – since connectors vary in diameter, stepped adaptors are available. The unit should also have a mains power socket to enable you to plug in a power tool; when you switch on the tool the extractor is activated, and it continues to run on for a few seconds after the tool is turned off.

The hose of a vacuum extractor is connected to a power tool.

Workshop extractors

For small- to medium-size workshops, a mobile extractor will collect large quantities of waste from machines such as planer thicknessers (jointer planers) or table saws. To cope with this, the hose diameter is larger than on smaller vacuum extractors and waste is filtered through an upper fabric bag and deposited in a disposable plastic sack, which can be easily removed from the machine. Make sure you choose the correct type of extractor to suit your woodworking needs, because some units are designed solely for chip collection, while others have specific filters for coarse or fine dust.

For extraction from several machines it is better to install plastic or galvanized steel ducting along the workshop walls or ceiling. Sections of pipe are clipped together, with bends where necessary, to form a complete run, and each machine is linked to the system, with waste fed directly back to one extractor. There is a danger of static electricity build-up with PVC pipes, so make sure the ducting is earthed correctly to prevent sawdust igniting.

A mobile dust extractor can be moved around the workshop.

Filtration

A certain amount of fine dust will always linger in the air after machinery or tools have stopped. These airborne dust particles are too small to be collected by a normal extractor, but can be removed by installing a simple filtration unit. This is either suspended from the ceiling or mounted on a shelf, and is designed to run as required and then shut down automatically after you have left the workshop.

A dust filtration unit removes airborne dust.

Dust masks

Disposable face masks are relatively cheap and will prevent dust causing throat irritation. Masks are rated according to dust particulate size, and in Europe a P2 mask is best for general woodworking. If no code is visible on the mask you should avoid buying, because it may not comply with the correct safety standard. For greater protection use a more elaborate respirator mask with disposable filter, while dual-cartridge respirator masks are safer when spraying lacquers and similar finishes.

Battery-powered respirator mask.

safer woodwork

A workshop may seem a pleasant environment in which to spend a relaxing few hours, but it is not without its dangers – using sharp hand tools and machines demands your full concentration. Never be tempted to remove the guard from a saw or planer (jointer) to save a few minutes on a task; find an alternative, safer way to machine that groove or cut that profile. Then there is the potential fire hazard of working with timber (lumber) and finishes, not to mention the fine dust that lingers in the air. Learn to control the risks so that potential injury to yourself or a visitor is minimal. Be even more aware of safety issues if children are present.

Tools and machinery

Unplug power tools before changing a blade or cutter, and use a pair of work gloves if handling circular saw or bandsaw blades. Planer (jointer) knives are notoriously sharp, so work gloves are vital when you are replacing these.

Get into the habit of plugging power tools into an RCD (residual current device). If you should accidentally slice the cable, this isolates the power supply within a millisecond to prevent an electric shock.

Honed plane blades and chisels can cause nasty cuts, but are more likely to slip if their edges are dull. Correct use of these tools will help you to reduce risks.

When not in use keep chisels in a tool roll or stored in a rack, or slide plastic blade guards over the exposed ends. Keep both hands behind the cutting edge when using chisels.

Eyes, ears and feet

Routers, planers (jointers) and circular saws eject chips as soon as their blades touch the timber (lumber), so always wear safety glasses or a visor to prevent debris getting in your eyes.

Toughened sunglasses are available if you are working outdoors in bright daylight.

For woodturners a better solution is the powered respirator, which completely protects your face. The visor does not mist over because filtered air is passed over your face via a battery-powered fan. These units can be worn for several hours before recharging is necessary and also give excellent protection when operating a router.

Although hearing loss is gradual and may not initially seem to be a problem, ear protection is vital – especially when using power tools or machines for more than a few minutes at a time. Wear disposable foam plugs, which can be bought in quantities

Related info
Layout and workflow (see page 12)
Dust control (see page 20)
Safety (see page 92)

to reduce the cost. Alternatively invest in a pair of ear defenders and hang them alongside the table saw or planer (jointer) so you are prompted to wear them.

Your feet may not seem as much at risk as face or hands, but drop a length of timber (lumber) on unprotected toes and you will soon know about it. If you cannot bring yourself to wear boots or trainers with steel toecaps, try a pair of durable leather shoes instead. It may be tempting in the heat of summer to wear open-toe sandals in the workshop, but a chisel rolling off the bench can do a lot of damage.

Fire and chemicals

Most woodworking shops are full of potential fire hazards, so it is essential to control the risks. Fit a fire extinguisher close to the door and make sure it is regularly checked. If in doubt about whether to get a powder, foam or water extinguisher consult a fire safety expert – in an emergency, using the wrong type could make a fire worse. If your workshop is part of the house or garage, make sure your insurance cover is still valid if there should be a fire. Install a ceiling-mounted smoke alarm and test it frequently and restrict smoking to outside the workshop.

It is not just wood or sawdust that can lead to fire, but also chemicals such as varnishes, polishes and solvents. These should be stored in a steel cabinet, while cloths used for applying finishes should be disposed of by unfolding and laying them flat to dry out outdoors – they could ignite if screwed up and tossed in a bin. Wear thin latex gloves when using oils or stains, which can be thrown away afterwards. Don't use a specific finishing product without sufficient ventilation, if necessary. In good weather consider finishing outdoors, though this does make the process harder to control. When using any finishing product, check the manufacturer's instructions on the tin or bottle first. Although many chemicals are now much safer to use there may still be a risk from paint removers, which are sometimes required when preparing recycled timber (lumber). Wear eye protection and sturdy gloves to prevent any splashes harming your skin. If you do get a burn, wash the skin under cold running water for several minutes then cover the injury with a sterile dressing. If the injury is bad enough to need a hospital visit, take details of the offending chemical with you.

First aid

No matter how small your workshop, keep a first aid kit handy and make sure it is readily visible even to visitors. Check the contents regularly, topping up any missing or out-of-date items.
The kit should contain:
tweezers
alcohol cleaning pads
eye wash solution
plasters (adhesive bandages) and dressings
scissors
safety pins
sticky (adhesive) tape

hand tools

Virtually every woodworker needs a few basic hand tool skills, such as measuring and marking out. Some of these require more practice than others, but whether you will be turning bowls, making doors and windows, or creating fine musical instruments, the same basic techniques apply. Even a professional woodworker who relies on power tools and machines occasionally still has to use hand tools. Buy the best quality you can afford, even if this means obtaining tools secondhand – with a little effort it's often possible to rejuvenate that old jack plane or chisel until it performs as sweetly as the day it was made.

basic toolkit

Some hand tools are essential, even if you plan mainly to use machines for your woodwork. Don't be tempted to buy a huge toolkit from the start, even if you can afford to. Only buy tools as you need them, focusing on quality rather than quantity – many professional woodworkers do most of their work with surprisingly few tools. Remember that a solid workbench is probably the most important tool in many workshops.

Measuring and marking

▲ **Pencils** A 2H grade pencil will give a precise line. For most woodwork avoid the carpenter's pencil, which is much too heavy for accurate marking.

▲ **Tape measure** Useful when measuring sheet materials in the workshop or timber (lumber) at the sawmill. Long tapes are bulky: short ones tend to have narrow blades. A good length is 5m (16ft).

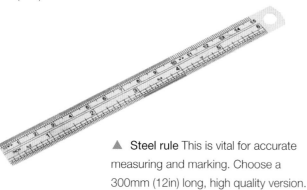

▲ **Steel rule** This is vital for accurate measuring and marking. Choose a 300mm (12in) long, high quality version.

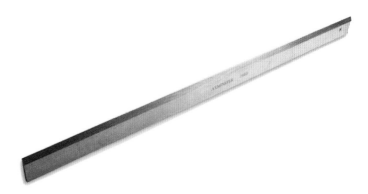

▲ **Steel straightedge** An engineering tool, use this for drawing, slicing veneer and checking timber (lumber) during planing. A hole at one end enables you to store it on a hook on the wall.

▲ **Try square** One with a blade about 230mm (9in) long is a good choice, and if it has a brass facing on the stock it will be more durable. A quality combination square, with its adjustable blade, may be more versatile and for finer woodwork a small engineer's square is perfect.

▶ **Sliding bevel** This can be locked in position at any angle from 0 to 180 degrees.

> **Related info**
>
> Workshop storage (see page 18)
> Location and planning (see page 10)
> Power tools (see pages 60–89)
> Woodworking machines (see pages 90–119)

Planes

▲ **Jack plane** Used for preparing sawn timber (lumber) or planing edges of manufactured boards, this bench plane is a good all-rounder.

▲ **Block plane** This is ideal for cleaning up end grain or for adding chamfers.

▲ **Marking knife** The tip is used to mark or scribe joints, particularly shoulder lines. Some knives are ground on one side only, others on both.

▲ **Marking gauge** It is worth paying a little more for a tool with brass strips set into the hardwood stock – cheaper tools without this feature will wear faster.

▲ **Dividers** Used at the drawing board when working out proportions and at the bench to copy measurements. They are also used extensively in woodturning to measure the diameter of the component at the lathe.

▲ **Compasses** Used to draw circles and curves. Also useful for scribing cabinets and trim to match uneven walls.

Quality tools

Keep your tools in good condition by cleaning them after use, protecting blades with edge guards and storing each item correctly.

Chisels

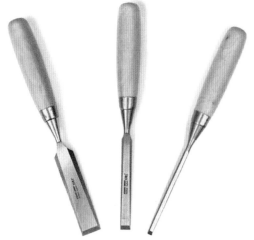

▲ **Bevel edge chisels** Start with a set of four: 6mm (¼in), 12mm (½in), 19mm (¾in) and 25mm (1in). The handles should be comfortable, whether traditional hardwood or plastic.

Saws

▲ **Tenon saw** Essential when cutting joints – choose a blade length of 250mm (10in) or a large dovetail saw instead.

▲ **Handsaw** For sawing timber (lumber) and sheet materials. For most woodwork, crosscut teeth are the best option.

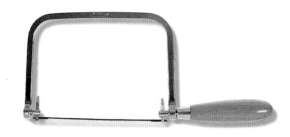

▲ **Coping saw** Using a coping saw is the easiest way to cut curves in most materials.

Common tools

▲ **Mallet** Used to strike chisels and assemble joints – make one as a simple project.

▲ **Hammers** Choose a cross pein hammer, and for carpentry work you will need a heavier claw hammer.

▲ **Oil, Japanese water or diamond stones** These are for sharpening plane blades and chisels.

◀ **Honing guide** This is used when sharpening to get a perfect edge to blades and plane irons.

▲ **Rasps** Used for fast wood shaping – specialist spokeshaves can follow later.

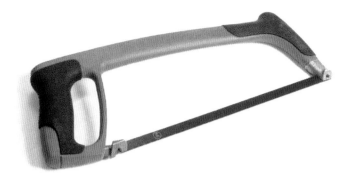

Hacksaw You may need to cut metal as well as wood, and this is ideal. Saws have standard blades 300mm (12in) in length.

Cramps (clamps) For holding work to the bench and gluing up. Quick-action cramps (clamps) are versatile and good value, while traditional G cramps (C clamps) are sturdier.

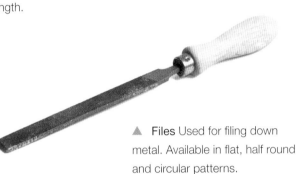

Files Used for filing down metal. Available in flat, half round and circular patterns.

Centre punch To mark hole centres accurately before drilling.

Pin hammer For driving in panel pins.

Nail punches (nail sets) To set nail heads beneath the timber (lumber) surface before applying filler.

Cork sanding block Wrap abrasive paper around this for hand sanding.

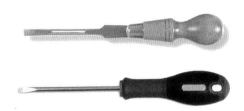

Screwdrivers Slotted and Pozidriv tools in at least two sizes of each pattern.

Cabinet scrapers A scraper is often the best tool to clean up timber (lumber) with wild grain and prevent tear-out.

Power tools

◀ **Cordless drill** This almost makes the hand drill redundant for drilling holes, and is more convenient than a 240V tool. Choose lip-and-spur bits for wood, although flat bits are faster for bigger holes where accuracy is not vital.

◀ **Jigsaw** For cutting sheet materials to size and for sawing solid wood.

◀ **Router** A compact ¼in collet tool is almost essential when making joints, moulding edges and so on. Buy a small set of router cutters at the same time.

◀ **Random orbit sander** Replacing hours of tiresome hand sanding, this tool will produce a great finish when it is fitted with the appropriate abrasives.

Secondary toolkit

As your skills develop you will find that a few extra tools to supplement the basic toolkit will make life easier in the workshop.

▲ **Mortise gauge** Whereas a marking gauge has only one pin, this tool has three making it more versatile.

▲ **Cutting gauge** Instead of a pin this tool has a knife blade and is handy for marking dovetail shoulders.

▲ **Spokeshaves** With convex and concave soles for creating curves.

▲ **Gents saw** Like a small dovetail saw, this tool is used for finer cuts.

▲ **Mitre square** Like a try square, but with the blade set at 45 degrees.

▶ **Sash cramps (clamps)** For gluing panels and boards together – best to buy them in pairs.

◀ **Smoothing plane** Smaller than a jack plane, for finishing wood.

how to buy tools

Depending on whether you want to buy brand new workshop tools or are happy to settle for second-hand, there are plenty of buying options. From car boot (yard) sales to high-tech Internet stores, each has its advantages. You are no longer restricted to fixed store opening hours either, since most mail order companies offer next day delivery for tools.

Ancient or modern?

Gleaming new tools may be very tempting but sometimes you can be better off buying old ones, which are often much better value for money. Not only are they usually cheaper than new tools, but also the steel quality can be better in old tools than in modern alternatives – especially with chisels. However, a hand tool revival means there is an array of new hand tools on the market that are of superb quality – with matching price tags. Some of these tools are based on traditional designs, while others are innovative creations.

You can often find perfectly good second-hand tools at car boot (yard) sales or in junk shops, but do your homework first – you could pick up a bargain or pay through the nose for a tool of uncertain quality. Do not be tempted to buy a power tool at such an event; it may well seem cheap, but there is no guarantee that it works and it could even be dangerous. During the summer months timber (lumber) and woodland craft festivals are often held around the country and these are another source of old hand tools. You will often find boxes of saws, chisels and planes in need of a good home and most of these events will also have green timber (lumber) for sale from locally felled trees.

> **Related info**
>
> Basic toolkit (see page 26)
> Secondary toolkit (see page 30)

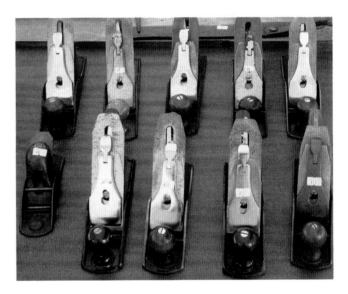

Specialist shops are a good source of second-hand tools, which are often of much better quality than new equivalents. Old wooden planes, in particular, are plentiful and cheap.

Tool stores and DIY centres

Small, specialist tool stores offer plenty of advantages to the woodworker, although they cannot always compete with the bigger DIY centres on price and product choice. Their knowledge of tools is usually second to none, and after-sales service is more important. They are more likely to stock that spare part or hand tool you thought was no longer available, and if a tool is not on display ordering it for you is usually not a problem.

DIY centres often carry a wide range of hand and power tools and are good places to see what is new on the market. Most have their own brand of budget power tools, which are good value for occasional jobs. DIY centres also open at evenings and weekends, which is a service that specialist tool shops normally cannot offer.

Internet buying

With Internet access you can check websites to research the newest power tool or must-have gadget, from anywhere in the world. In our global economy woodworkers living in Europe can buy from stores in the USA, and vice versa. Internet buying usually means fast delivery if the items you want are in stock, and mailing to remote destinations is fairly straightforward. Retail shops must compete with on-line sellers to survive, so prices are generally competitive either way. However, comparing tool prices is easier if you have a clear idea of what you need – and larger tool suppliers often have their own printed catalogues, which can be less tiring to read than scrolling through on a computer screen. If a tool is faulty or of suspect quality, it is also generally quicker to return it to a high street store than to mail it back to an online or mail order supplier.

Woodworking shows

Woodworking shows offer the chance to see tools and machines being demonstrated, and you can talk to tool specialists and get the opinion of demonstrators. Often a show will also be the only opportunity to try a tool before buying. Aimed at woodworkers of all skill levels, these are the places to pick up a bargain. Most exhibitors are keen to return to base with an empty truck, so time your visit right and you may be able to knock down the price of that machine you have had your eye on. For major shows or events in your area it pays to check magazines and the Internet.

Woodworking magazines

Most woodworking magazines encourage readers to advertise unwanted tools and machinery for sale. Such items are often well cared for and in excellent condition – although good quality, well-priced tools do tend to sell fast. Woodworkers are generally an honest crowd, but take the usual precautions before paying; ask a seller to email a photo of the tool you are interested in buying, so you can assess its condition before making a decision.

Buying tools

1. With hand tools, decide first on new or old – both offer advantages. New tools should have all necessary components, while an old tool may have parts missing.

2. With new tools it is easier to get just what you want. Talking to other woodworkers and research on the Internet will help you come to an informed decision. If budget is tight, it is also easier to compare prices on new tools.

3. Don't expect a brand new hand tool to work properly straight out of the box – many will need tuning (fettling) to work efficiently. This is particularly true of bench planes.

4. Buy the best you can afford, particularly with hand tools. A cheap tool may be frustrating to use or result in poor work. A more expensive, better quality tool will be more reliable and last longer.

5. If buying old hand tools, examine them carefully first. Check that wooden plane bodies or handles are not cracked and do not contain worm holes.

6. If choosing between a hand or power tool for a certain task, think about how much time you have available. Power tools are often faster and more efficient, but are noisy and create far more mess than hand tools. If time is not an issue, hand tools are quieter and can be more satisfying to use.

measuring and marking out

Most woodwork relies on accurate measuring and marking of the timber (lumber) first and foremost; this establishes the standard of work that follows. Take great care at this stage, checking and rechecking dimensions before you start to cut – get it wrong and the rest of your project will suffer. The same measuring tools are used whether you are working with machines or by hand. Europe uses the metric system for measuring, while North America is largely imperial – many tools are now marked in both.

Related info

Setting out (see page 164)
Cutting lists (see page 157)
Planing primary surfaces (see page 166)
Construction methods (see pages 178–213)

Marking timber (lumber)

The basic tool for marking out is the pencil, which comes in various grades from soft to very hard. For accurate marking of joints a 2H grade is best – you will need to resharpen a common HB (medium) pencil often, because it is much too soft for most woodwork. Avoid using the carpenter's pencil, with its heavy lead, for fine woodwork – although it is fine for identifying sawn planks.

Many craftsmen favour a marking knife for setting out joints, since it produces a finer line than a pencil. When buying check steel quality, because the blade needs to withstand being drawn against try square blades. Sharpening is carried out as for a chisel. To use a marking knife, hold firmly against a rule or square, perpendicular to the surface, and strike a line with the tip. You can mark precise lines because the blade is bevelled on one side, which should face towards waste material. Knife lines will remain even after assembly is finished, so plan ahead and mark selectively to avoid unwanted lines showing on the finished item.

Measuring timber (lumber)

For measuring long boards or marking out sheet materials a retractable tape is essential, although it tends to be less accurate than a long steel rule over a distance. A 5m (16ft) long tape will be fine for most jobs. You can lock the blade open with a button, while the loose end hook compensates for both internal and external measurements. The steel rule is undoubtedly a more accurate measuring tool – the 300mm (12in) size is preferable in the workshop, a pocket-size 150mm (6in) rule is useful for checking timber (lumber) sizes during machining and a 600mm (24in) version is indispensable for producing drawings. Rules are normally etched in 1mm ($\frac{1}{32}$in) increments, since 0.5mm ($\frac{1}{64}$in) graduations are difficult to see. Every workshop should also have a steel straightedge – 1000mm (36in) is a useful length. Use this tool to check timber (lumber) for flatness when planing, and also tools such as bench planes.

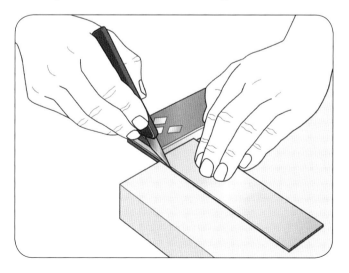

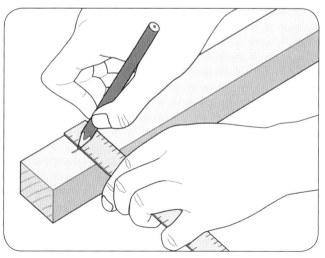

Try squares and sliding bevels

A crucial tool in any workshop, the square is used for marking out at 90 degrees to an edge, as well as checking the accuracy of corners and joints. Although you can buy cheap plastic squares, for precision work the steel engineer's square is the best option. With its brass-faced rosewood stock and steel blade, the traditional try square is often decorative as well as functional. Squares are classified by blade length, with a 230mm (9in) version a good choice for most woodwork. If marking out complex joints, you may find a smaller engineer's square particularly useful.

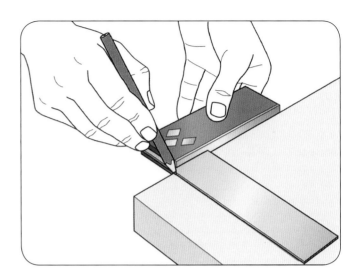

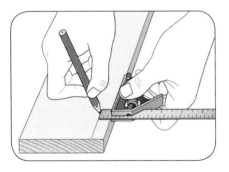

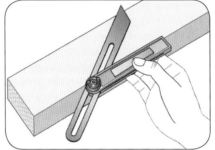

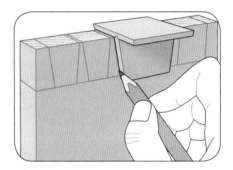

A combination square is more versatile than a standard try square, although it is available in fewer sizes. One edge of the stock gives 45-degree mitres, while you can slide the graduated blade to measure router cutter projection, grooves or rebate (rabbet) depths. However, a fixed mitre square is useful for the regular marking out of 45-degree angles.

For marking out or checking any angle that does not correspond with a regular square, you will need a sliding bevel. This is simple to use; you either lock its adjustable blade with a lever or with a screwdriver.

Dovetail joints need to be marked out at certain angles (1:6 for softwood, 1:8 for hardwood) to be effective. Although you can do this with a sliding bevel, a dedicated template is faster; some incorporate both angles on the same tool.

Checking a square

If your square is not precisely 90 degrees, all subsequent work will be inaccurate, from initial marking out to assembling carcasses. It is easy to check accuracy by holding the stock against a dead straight edge – either wide timber (lumber) or sheet material – then with a fine pencil draw a line down the outer edge of the blade. Turn the square over and repeat the process; if the lines coincide exactly the tool is accurate. Always check a square afterwards if you should drop it. You can file the outer blade edge to regain accuracy, but it is almost impossible to do this on the inside.

Marking and mortise gauges

Gauges are simple tools used for scribing lines parallel to an edge. When fitted with a sharp pin, a marking gauge marks along the grain. A cutting gauge is fitted with a small scribing knife instead of a pin, and is used to scribe across the grain. Sharpen this knife with a single bevel, like a marking knife, and when using it keep the bevel on the waste side of the line. You may choose to have both types of gauge, but a standard marking gauge can be fettled to work well both with the grain and across it. To maintain a fine line you will need to file the pin occasionally.

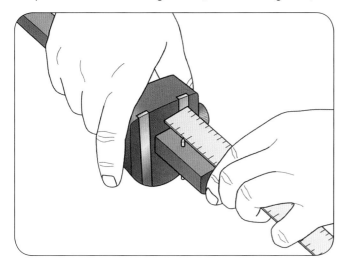

1 Set the gauge to width using a steel rule, then tighten the thumbscrew. Check the setting and if adjustment is necessary, slacken off slightly and tap the stem on the bench. When the measurement is correct, retighten the adjuster screw. Adjust a cutting gauge in the same way.

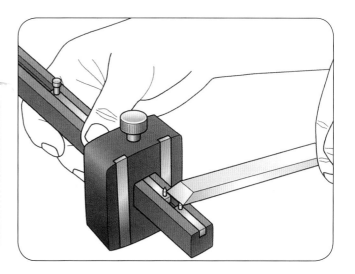

2 When using a gauge, push it away from you maintaining sideways pressure on the stock to keep it tight to the edge of the timber (lumber).

Buying squares

Always check for accuracy when buying a try square. Take along a wide piece of MDF or timber (lumber) with a dead straight edge, plus a sharp pencil. Check as explained on page 34.

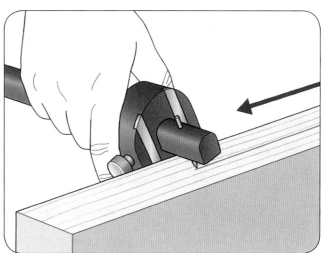

3 A mortise gauge is fitted with a pair of pins and is used to scribe the parallel lines needed when marking out mortise and tenon joints. One pin is fixed; the other is adjusted to the chisel width being used for the mortise. If you envisage making lots of joints, choose a gauge on which the sliding pin can be moved by a screw thread adjuster.

cramps (clamps)

Cramps (clamps) are an essential part of any toolkit. Whether securing a plank to the bench when routing or assembling a cabinet, the variety of cramps (clamps) can make it hard to choose the most suitable for your work – most woodworkers have several types for different tasks, so buy cramps (clamps) as you need them and build up a good selection gradually. When gluing up a project that requires a lot of cramps (clamps), always do a dry run first to check you have enough; if not, you may have to rethink your approach.

Quick-action/solo cramp (clamp)

Popular quick-action cramps (clamps) are light in construction but can exert a great deal of pressure. Plastic or reinforced resin jaws are mounted on a steel bar, and pulling the trigger moves them together. On some types, reversing the jaws enables you to force components apart when doing a dry assembly. Quick-action cramps (clamps) are generally designed for single-handed use.

The solo cramp (clamp) is operated in the same way as a sealant gun, so you can position and tighten it with one hand. Its steel frame has nylon pads for protecting the workpiece and it is ideal for holding down timber (lumber) on a portable workbench.

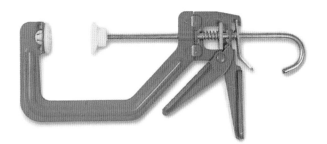

Related info

Adhesives and assembly (see pages 214–223)
Construction methods (see pages 178–213)
Workshop storage (see page 18)

G cramp (C clamp)

The traditional G cramp (C clamp) is a useful general purpose tool, perfect for exerting tremendous pressure when necessary. Normally made from malleable cast iron they are virtually indestructible, but in some situations they can be awkward to tighten because you need both hands to position them. Depth capacity is up to 250mm (10in) on the bigger cramps (clamps), though much smaller sizes are available. For effective pressure away from an edge, use a deep-throat G cramp (C clamp).

A clever variation is the edge cramp (clamp), which is useful for adding solid wood lipping to the edge of sheet material. Three screw adjusters increase the versatility of this specialist cramp (clamp).

F cramp (clamp)

Because these can be adjusted rapidly they are also called speed cramps (clamps). Time is often crucial with many intricate gluing tasks, so these are a favourite with many woodworkers. They are easier to position than G cramps (C clamps), and shoes tend to be reinforced plastic or cast metal; you slide the shoe along the steel bar and tighten by rotating the handle. The bar's rear edge is often serrated so the shoe locks as you tighten up. From heavy to lightweight, these are excellent general purpose tools with capacities as much as 1500mm (60in) on professional models.

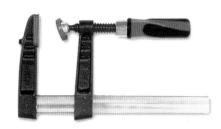

Cam cramp (clamp)

Popular with musical instrument makers, lightweight cam cramps (clamps) are perfect for applying pressure gently without harming delicate wooden components. The heads are usually hornbeam or beech, faced with cork to prevent damage, and fitted to a rectangular steel bar. You can make your own cam cramps (clamps) relatively easily.

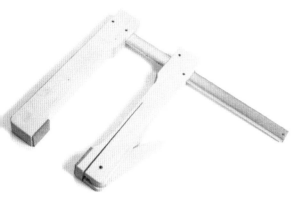

Spring cramp (clamp)

The spring cramp (clamp) is the simplest cramping (clamping) device, particularly useful for gluing small items that do not need much pressure, such as when model making. It can be plastic, although heavier versions are steel. The heavier spring cramps (clamps) can be quite powerful and are handy for holding routing templates or guide battens. Ratchet action is a feature on more sophisticated versions, making them faster to tighten and release.

Mitre cramp (clamp)

Gluing two pieces of mitred wood together at 90 degrees can be tricky, especially if you need to reinforce the joint with pins, but the cast iron mitre cramp (clamp) enables you to assemble mitres cut at 45 degrees easily. Some types have a slot to enable you to saw the mitre before gluing together. This type of cramp (clamp) is used primarily for picture framing.

Band cramp (clamp)

Picture frames with unusual shapes, or chairs, can be particularly awkward to cramp (clamp) up and it is often not possible to use a sash cramp (clamp). The band cramp (clamp) incorporates four plastic or alloy jaws plus a nylon or steel band that passes through a tensioner device. Simply pass the band around the structure and tighten the knob.

Sash cramp (clamp)

With their large capacity – up to about 1650mm (65in) – the sash or bar cramp (clamp) is particularly useful when gluing boards together edge to edge, or assembling joinery items such as windows or doors. It consists of a steel bar drilled with a series of holes, with a shoe at one end that is tightened using a tommy bar. You slide the opposite shoe along the bar, inserting a metal pin in the appropriate hole to prevent it moving. With heavy T-bar cramps (clamps) you can apply enormous pressure, although lighter duty cramps (clamps) may have extruded aluminium bars that are not so strong. You can make an effective sash cramp (clamp) simply by buying a pair of steel shoes and preparing a piece of hardwood cut to size. You are limited in capacity only by the length of the timber (lumber), but remember to space out the pin holes equally.

Pipe cramp (clamp)

Similar to sash cramps (clamps) – but without pins – pipe cramps (clamps) use off-the-shelf steel plumbing pipes as bars. Cut your own thread or buy them pre-threaded, then screw on a fixed head that includes a handle. With the pipe sawn to the required length, slide the second head or tailstock onto the far end. Instead of a steel pin, this head is locked via a clutch system. Since pipes can be swapped, you only need buy a couple of pairs of cramping heads to get started. Pipe diameters are either ½in or ¾in, and can be joined to extend their capacity.

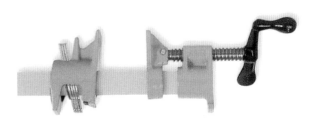

Cramp (clamp) care

Try to wipe off excess glue from cramps (clamps) before it dries – although if this is not possible dried glue can usually be chipped off. If you have an unheated workshop any spray steel cramps (clamps) occasionally with an all-purpose lubricant to prevent them from rusting, and lightly oil all the screw threads.

saws

Although any saw will cut wood, it is the kind of tool and size of its teeth that categorize it for a particular task. There are different saws to cut delicate dovetail joints, rip thick planks to width or cut thin veneer. They are not restricted to straight cutting; curved cuts are possible with frame saws. Teeth on Western saws cut on the push stroke, while the increasingly popular Japanese tools cut on the pull stroke.

Back saws

Unlike a handsaw, the back saw has a strip of brass or steel folded along the top edge of the blade to stiffen it and increase weight. Dovetail saws are shorter, with smaller teeth for finer cutting. For general cutting and larger joints a tenon saw is more popular. Hardpoint versions of the popular sizes of tenon saw are available, although traditional back saws will have resharpenable teeth. They are fitted with closed or open hardwood handles, which can be decorative but above all should be comfortable. The blade on a tenon saw varies between 250 and 350mm (10–14in) in length, with teeth size 12–16tpi (teeth per inch). A dovetail saw blade is 200–250mm (8–10in), with teeth as fine as 20tpi – cabinetmakers find a 200mm (8in) dovetail saw handy. Even smaller, the gents saw blade is 100–250mm (4–10in) with teeth around 20tpi; use this for precision sawing. The smallest member of the family is the jeweller's saw. To cover a wide range of cutting tasks, you will find that many craftsmen use two or three different back saws.

Dovetail saw (200mm/8in blade).

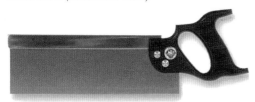

Tenon saw (300mm/12in blade).

Gents saw (150mm/6in blade).

Related info

Sharpening saws (see page 53)
Cordless power tools (see page 62)
Portable saws (see page 68)
Woodworking machines (see pages 90–119)

Handsaws

The handsaw is grandfather of the saw family, used for ripping and crosscutting timber (lumber), as well as sawing manufactured boards. Teeth shape and size determine the tool's function: a crosscut saw has 7–8tpi, a rip saw is coarser at 4–5tpi, and finest is the 10–12tpi panel saw, which is ideal for cutting boards and joints. Handsaws are manufactured in several sizes, from around 400 to 660mm (16–26in). Traditional saws have resharpenable teeth, though you cannot sharpen hardpoint saws. Expensive handsaws may feature a taper-ground blade that varies in thickness, making the saw easier to use. On large saws teeth traditionally had a slight convex curve instead of lining up in a straight row, which made the sawing action more efficient, although this is a feature rarely seen on new saws.

Rip saw (660mm/26in blade).

Handles on saws are usually lacquered hardwood, riveted or screwed to the steel blade. Top quality saws may have cherry, walnut or maple handles, though beech is usual. Cheaper plastic handles are standard on hardpoint saws, since ergonomics often dictates a textured soft grip. For cutting sheet materials efficiently choose a hardpoint saw: chipboard and MDF can blunt teeth rapidly, but a hardpoint's heat-treated teeth will stay sharp much longer than a traditional saw. A good size for general purpose sawing is 550mm (22in).

Hardpoint saw with teflon-coated blade.

Curved cutting

The frame saw is used specifically for cutting curves in timber (lumber); you can swivel its narrow blade through 360 degrees. Suspended in a steel frame, the blade is tensioned by tightening the threaded handle or flexing the tool and locking it with turn buttons. The coping saw is the most popular tool for curved cutting; its blade slots onto retaining pins on the frame, which you remove by unscrewing the handle. Standard blade length is 150mm (6in), with about 14tpi. The smaller piercing saw is used for marquetry, model making and even soft metals; it has a finer blade of 130mm (5in).

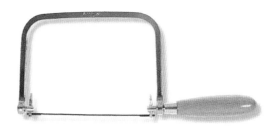

Special saws

A few saws are designed for specific cutting tasks and may not be used regularly. Keyhole and compass saws, with their long, pointed blades, will cut tight circles – drill a hole in the wood before inserting the tip. A veneer saw blade has a convex row of teeth for straight cutting and is used with a straightedge for very fine cuts. The thin, flexible blade of the flush-cutting saw is designed to remove wooden plugs level with the wood – teeth will not ruin a surface because they are set only on one side.

Use a flush-cutting saw to trim close to the surface.

Japanese saws

Japanese saws have incredibly sharp teeth, producing a very fine kerf. They cut on the pull stroke, unlike Western saws that cut as you push them, which means blades are under tension and are much thinner. Many craftsmen prefer the cutting technique associated with a Japanese saw, although it does not suit everyone. You can follow a pencil line more easily, with the razor-sharp teeth cutting quickly and efficiently. Ryoba and kataba saws (rip and crosscut) have thin blades that pass through the wood, while dozuki saws (tenon) have rigid steel backs. Handles are traditionally covered in bamboo, although contemporary versions have plastic handles. Special feather files enable you to sharpen the teeth, although the harder steel used does mean that these saws stay sharp longer than equivalent Western tools. Many have replaceable blades so are not resharpened.

Mitre saws

A favourite tool of picture framers, the mitre saw can cut precise angles (mitres) in timber (lumber). You can also saw pieces to preset lengths at virtually any angle, rather like using a frame saw held in a jig. Rotate the blade to cut up to 45 degrees left or right, with stops at these common angles and at 90 degrees for easy setting – a protractor scale acts as a visual guide when locking the saw at other angles. Timber (lumber) is placed on the base and held against a rear fence, which may include a cramping device. Set the cutting depth with stops on the blade.

Saw teeth

You can discover how coarsely a saw will cut by examining the size and number of teeth on its blade. Teeth size is given in either tpi (teeth per inch or 25mm) or ppi (points per inch or 25mm). To establish the tpi, measure the number of complete teeth per inch (25mm) along the base of the teeth, but for ppi count the number of points per inch (25mm). This means that on the same blade there is one ppi more than tpi, so 10ppi is the same size as 9tpi.

Teeth on rip and crosscut saws are shaped slightly differently; the front edge of a rip tooth is vertical and the tooth is filed at 90 degrees to its face – rip saw teeth operate like small chisels, chopping waste away as they move along the grain of the wood. A rip saw is normally better suited for cutting dry hardwoods.

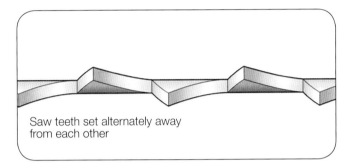

Saw teeth set alternately away
from each other

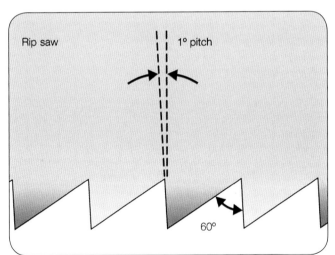

Rip saw — 1° pitch

60°

Recycling old saws

When the teeth on a hardpoint saw become too blunt for woodwork, relegate it for use in the garden instead.

A crosscut tooth slopes away from vertical and is filed at an angle – crosscut teeth work like tiny knives to sever the wood fibres. A crosscut saw is usually better for cutting damp softwoods. There is no hard and fast rule when it comes to cutting, however – some craftsmen reset their panel saws (fine teeth) with a rip tooth pattern instead.

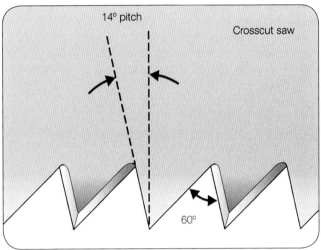

14° pitch — Crosscut saw

60°

Hardpoint teeth tend to follow either a universal or triple-ground pattern. Triple-ground teeth are similar to those on Japanese saws and cut on both pull and push strokes. Since they are hardened by an electronic process, hardpoint teeth cannot be resharpened.

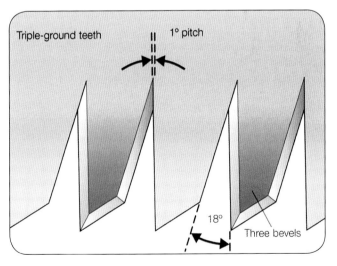

Triple-ground teeth — 1° pitch

18°

Three bevels

planes

One of the fundamental woodworking skills is planing timber (lumber) to size. Once wood is sawn to length it is then usually prepared to exact dimensions and surfaces are planed completely straight and flat, making sure edges are square. This means frequent checking with a try square and straightedge. Using a finely tuned plane to produce a wafer-thin shaving is one of the most satisfying woodworking experiences.

Related info
Sharpening hand tools (see page 52)
Cordless power tools (see page 62)
Power planers (see page 76)
Planers and thicknessers (see page 108)
Planing primary surfaces (see page 166)

Bench planes

In its most basic form, a bench plane is used to convert timber (lumber) from its rough-sawn state into a board that is flat, straight and smooth. Depending on the size of the plane and how it is set up, you can trim a joint so it fits together snugly, or take a wafer-thin shaving from a surface in readiness for final finishing. The resulting finish from a sharp, finely tuned plane is generally better than the sanding process can ever produce. Premium planes occasionally have a bronze body, although most tools are made from cast iron. Traditionally both front and back handles are lacquered beech or rosewood, although cherry and bubinga are popular on newer, upmarket tools. You will find plastic handles on budget tools, although these can usually be upgraded with hardwood equivalents. Steel bench planes are produced in several sizes (see box The bench plane family on page 46) to suit different purposes. Shortest is the smoother, which produces fine shavings, followed by the longer jack and fore planes, which are useful all-round preparation tools. Longest is the try plane – also known as a jointer – for creating an accurate straight edge. Before machines took over the task of planing rough boards into dimensioned timber (lumber), a craftsman needed a full set of planes: a scrub plane for quickly reducing timber (lumber) to size, a jack or fore plane for preparing timber (lumber), a jointer plane for long edge joints, and finally a smoothing plane to tidy up that glued-up door, window or piece of furniture. Now woodworkers frequently use just jack and smoothing planes.

Adjusting the cut

A jack or fore plane should have a slight curve across the edge of its blade. It will then require minimum effort to produce a thin shaving that exits from the middle of the mouth. The shaving will be about 0.25mm ($\frac{1}{100}$in) thick at the centre, tapering to zero at the edges. A flatter blade creates a wider shaving, but demands greater effort and allow less control. A smoothing plane blade should have a straight edge, but with corners rounded off to prevent tramlines appearing in the timber (lumber) as you plane.

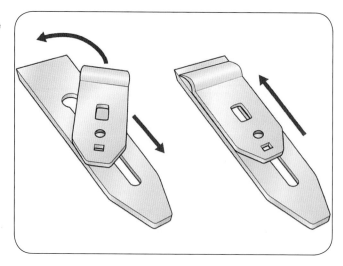

After sharpening a plane blade, place the back (cap) iron at 90 degrees to the blade when reassembling. Swivel the back iron round and slide forwards to prevent damage to the cutting edge.

Plane blades

Normally made from high quality carbon or tool steel, plane blades (or irons) fall into two categories. Standard bench planes are used with the bevel facing down, and the blade set in the tool at 45 degrees. A shaped back (cap) iron is screwed to the rear of the blade to guide the movement of the shaving through the plane's mouth and also prevent the shaving from tearing. For the plane to work effectively, make sure the back (cap) iron and blade seat together firmly. Block and shoulder planes have bevels face up, with blades set at a lower angle. The blade is locked to the body of the plane with a lever cap.

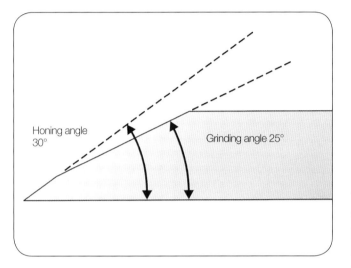

Honing angle 30°

Grinding angle 25°

Block planes

One of the most versatile hand tools, the block plane is used for trimming end grain or forming a chamfer. Used with one hand, you can also plane with the grain and it is ideal for narrow edges. The blade is seated at a low angle of 12–15 degrees in the body with the bevel uppermost; honed at 30 degrees, this gives an overall planing angle of 45 degrees, like most bench planes. More sophisticated tools have an adjustable mouth to reduce the opening in front of the blade, which enables you to take finer shavings when planing woods with irregular grain.

Specialist planes

Sometimes a woodworking project calls for a specialist plane, particularly when easing a joint, although these precision tools can be more awkward to sharpen and use than conventional bench planes. With its full-width blade the shoulder plane will cut tightly along a joint such as a rebate (rabbet), and held on its side, you can trim a tenon shoulder cleanly. Available in several body sizes, you will find one narrow shoulder plane is often adequate for most trimming tasks.

The rebate (rabbet) plane has an adjustable side fence and will cut clean, accurate rebates (rabbets). Although it has been largely replaced by the electric router, for small jobs it is often quicker and cleaner to use this tool than set up a power tool. The side rebate (rabbet) plane is a unique tool designed for tasks such as easing a groove in frame and panel construction.

Oriental planes

Simple Japanese and Chinese planes are favoured by many woodworkers and, with practice, these elegant tools can be satisfying to use. The technique used differs from their Western counterparts; instead of pushing the plane away from your body, you pull it towards you. Adjust the depth of cut by lightly tapping the thick blade with a small hammer. Oak is generally used for Japanese plane bodies, with rosewood and ebony found on Chinese tools. As well as smoothing planes, more specialized rebate (rabbet) and moulding planes are also produced.

Compared with their steel equivalents, wooden planes are simpler to set up and lighter to use. If you are unsure about investing, you can often buy old tools for next to nothing at car boot (yard) sales.

This contemporary Western smoothing plane has a cocobolo body.

The bench plane family

No matter what its make or age, every bench plane is designated by a standard number, which identifies overall length and blade width. Smoothing planes begin with the tiny and rare No.1, increasing in size to a No.4½.

Jack planes are numbered No.5 and No.5½.

Fore planes are designated No.6.

Try or jointer planes are numbered No.7 and No.8.

Most woodworkers will have a No.4 or No.5 plane as a minimum.

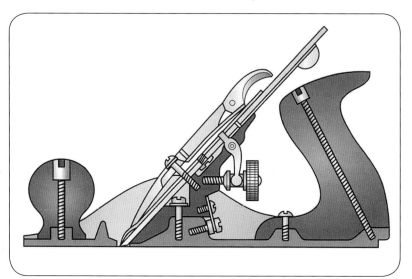

The components of a smoothing plane.

Choosing the right plane

Use a long plane to skim across the high points of an undulating surface, reducing these steadily until an entire shaving is produced. Avoid using a short plane on long boards, because it will follow the contours and not actually true up the surface.

chisels

A set of decent chisels is likely to become any woodworker's best friends. Made in several styles and sizes, most chisels are designed to be struck with a mallet so they need to be sturdy. To work efficiently it is important that cutting edges remain sharp, because then less effort is necessary for them to cut – which also makes them safer to use since there is less chance of them slipping. Learn how to sharpen chisels properly from the outset, using a honing guide if necessary.

Chisel patterns

Although you can buy chisels individually, a good quality set of four or five tools is a sound investment. The most popular blade sizes in a set are 6mm (¼in), 12mm (½in), 19mm (¾in) and 25mm (1in). Unless your woodworking demands more specialist blade widths, you may only ever need one or two extra chisels – for heavy-duty work you can buy chisels up to 38mm (1½in) in width. For efficient working a chisel should feel right when you pick it up, its handle matched perfectly to the weight of the blade. A large handle suits a wider tool, while a narrow chisel is easier to control with a small handle, and the comfort of the handle is as important as the quality of steel in the blade. Premium chisel handles tend to be boxwood, rosewood or hornbeam, while ash or beech are found on cheaper tools. Many modern chisels are fitted with plastic handles, such as polypropylene, which can be surprisingly comfortable.

Related info

Sharpening hand tools (see page 52)
Mortisers (see page 110)
Construction methods (see pages 178–213)
Turning tools (see page 116)

For carpentry and joinery work – rather than for much finer cabinetmaking – heavier, firmer chisels are more suitable. They are stronger than bevel-edge chisels and have much more substantial blades, although you cannot undercut with them in the same way.

Bevel-edge chisels are the best choice for use in the workshop. Both long edges have a bevel ground on them, so these tools are excellent for joint cutting, especially when cutting dovetails. Do not be tempted to chop mortises with them, though – since they have less steel in the blades they are less sturdy than other types of chisel and could snap.

For chopping mortises use a sturdier mortise chisel; its thick blade is built to lever out waste timber (lumber). To absorb shock when striking, the registered pattern of mortise chisel has a leather washer and steel hoop between handle and blade.

Japanese chisels

Japanese chisels are increasingly popular in Europe and North America – the blades are laminated from two grades of steel; the cutting edge is a thin layer of harder steel fused to a thick layer of softer metal. Western chisel blades are forged from one steel throughout, but the harder steel in an Oriental chisel means it will retain its edge much longer, when used correctly. Sharpen both types in the same way, although normally the Japanese tool has one bevel instead of two. The backs of Japanese chisels are hollow ground, so you can get the tool completely flat quickly, but eventually this hollow becomes visible across the edge as a result of continual honing. To restore the cutting edge, tap the back with a small hammer, which creates a new flat area for sharpening. Although European chisels tend to be lighter and faster to sharpen than Oriental equivalents, you can obtain delicate Japanese blades as narrow as 1.5mm (¹⁄₁₆in) in width. Handles on Oriental chisels are usually elegantly polished oak, with a steel hoop that allows them to be struck with a special hammer rather than a mallet.

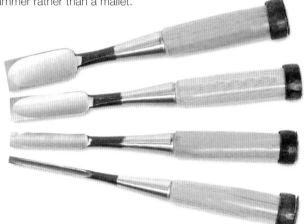

Using a chisel

Make sure the workpiece is cramped to the bench top or gripped in a vice before using a chisel. For safety, make sure you keep both hands behind the cutting edge, whether you are paring or chopping a mortise – don't ever hold timber (lumber) with one hand while using a chisel in the other. When you need to exert even more pressure during paring, it is time to hone the blade. In fact, it is not unusual to hone the edge every 30 minutes or so when using a chisel for any length of time.

Special chisels

For certain cutting tasks you may need more unusual tools. Paring chisels are perfect when making housing (dado) joints by hand – with their bevel-edge blades they are longer than other chisels and altogether more delicate, so should not be used with a mallet. They are available with blades from 6mm (¼in) to 32mm (1¼in) in width. Particularly useful for tidying up glued cabinets and wide joints is the cranked paring chisel; its handle is offset from the blade, which is used flat on the wood surface.

For cleaning really awkward corners you may need to use a skew-ground chisel. These are generally sold as a matched pair (left and right hand), though you could save money by regrinding two old chisels to 45 degrees. These tools are ideal for cleaning out the fibres when making the lapped dovetail joints that are used in drawer construction.

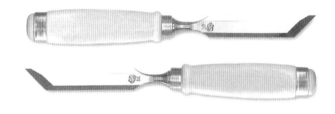

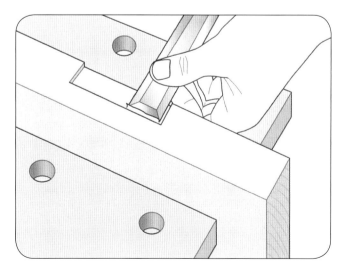

scrapers and spokeshaves

It may be the simplest tool in the workshop, but sharpened and used correctly the cabinet scraper will create a perfect, finished surface, while – rather like a small plane – the spokeshave is designed to cut convex or concave curves. Unlike a plane, the spokeshave is awkward to use initially, so practise on off-cuts before tackling that valuable piece of hardwood. Convex and concave curves are produced using different pattern spokeshaves, so it may be necessary to buy a pair for your work.

> **Related info**
>
> Sharpening hand tools (see page 52)
> Files and rasps (see page 54)

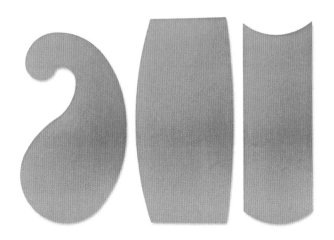

Curved and straight scrapers.

Cabinet scrapers

Most cabinet scrapers are rectangular with two straight cutting edges, although some are shaped with convex or concave curves. They are made from thin tempered steel and when sharpened correctly will create a fine shaving. If dust is produced instead, the tool needs resharpening.

Burnisher.

Using a cabinet scraper

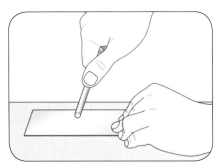

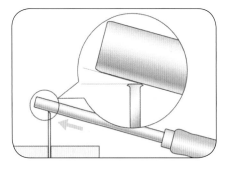

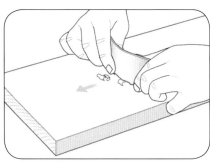

1 First draw a file along the edge to get it square, with the scraper held vertically in a vice. Then, holding the scraper flat on an oilstone, slide it to remove the burr. Turn the scraper over and repeat on the other face. Finally, with the scraper flat on a bench, draw a burnisher along the edge, keeping it flat.

2 Form a small burr on the cutting edges of the scraper by drawing the burnisher along it at a slight angle. Two or three firm strokes are all that is needed. Repeat the process on the opposite cutting edge, so you end up with a total of four burrs along the two long edges of the scraper.

3 Grip the scraper with both thumbs in the centre of the blade. With the tool leaning away from you, push it across the wood to make a cut. Experiment with the scraper angle and amount of pressure you apply to the blade.

Scraper plane

When using a cabinet scraper for any length of time your thumbs will get very hot because heat builds up from the friction involved. The scraper plane overcomes this problem and is a handy tool for cleaning up those wider boards and panels. It looks rather like a bench plane, but you sharpen the blade in the same way as a cabinet scraper. The tool has a flat sole and you can adjust depth of cut via a thumbscrew that applies pressure to the blade.

Spokeshaves

Premium quality spokeshaves tend to have hardwood handles and cast iron or bronze bodies, while cheaper tools are made from a single piece of grey iron. It is the shape of the sole that identifies the tool for the type of curve: for convex curves you need a spokeshave with a flat sole, while for concave curves a rounded sole is necessary. Except for the soles, the tools are the same; the blades are sharpened like a block plane. For chairmaking there are more unusual spokeshaves, used to shape legs, rails and seats.

There is no adjustment lever or screw on the simpler spokeshaves; to vary the depth of cut slacken off the cap iron thumbscrew and move the blade in or out manually. On more complex spokeshaves you can align the blade exactly, using screw adjusters to raise or lower the blade on each side.

Using a spokeshave

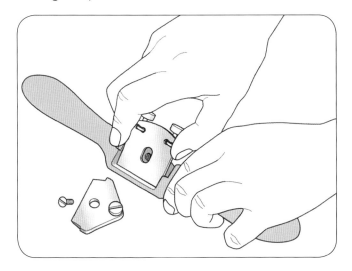

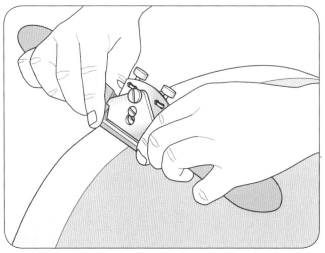

1 Once the blade has been sharpened, replace it in the tool with the bevel side down. If there are no separate adjuster screws, let the blade protrude slightly from the mouth and tighten the thumbscrew. Make sure the wood is held firmly in the vice. With the blade pointing away from you, grip the spokeshave with both hands, thumbs resting on the back of the handles.

2 With the tool on the wood, push it away from you using several short strokes. If adjusted correctly the blade should cut a thin shaving. If the shaving is too coarse or the tool is not cutting at all, readjust the blade. Work with the grain, changing direction if tear-out occurs.

sharpening hand tools

Sharp edge tools are not only much more efficient at cutting wood than dull tools; they are also safer because you will have greater control. It takes much more effort to push a plane along a board if the tool is blunt, while a tired chisel is more likely to slip and cause an injury than one that is razor sharp. When you feel more resistance when cutting or planing timber (lumber), it is time to resharpen. Get into the habit of sharpening the planes and chisels that you are likely to need at the start of each workshop session.

Related info

Planes (see page 44)
Chisels (see page 48)
Scrapers and spokeshaves (see page 50)
Saws (see page 40)
Sharpening a hollow chisel (see page 111)

Sharpening process

Chisels and plane blades (irons) are nearly always manufactured with their cutting edges ground at 25 degrees (primary bevel). For cutting wood cleanly and efficiently the steel must be further sharpened, or honed, to create a fine bevel of 30 degrees (secondary bevel). To do this you move the blade back and forth on a sharpening stone, having increased the angle by about 5 degrees. You can also resharpen some handsaws, unless the saw is a hardpoint tool. However, sharpening saws requires skill and patience to avoid damaging the tool so it is best to take them to a specialist dealer or saw doctor.

Primary bevel 25°

Secondary bevel 30°

Sharpening stones

There are several types of stone for tool sharpening, both natural and man-made. Almost every stone is used with a lubricant, which carries the steel particles created during honing – without this the stone would clog up as you move the blade along the surface. It is also feasible to use abrasive paper glued to a flat surface – silicon carbide paper is suitable, with water to lubricate.

Japanese waterstones are soft and wear rapidly, so they require frequent flattening on abrasive paper stuck to a flat surface. You can buy both natural and synthetic waterstones, which cut fast and come in grades from a coarse 800 grit up to an extremely fine 8000 grit. If buying a single stone choose a medium grade, around 1000 grit. Use finer stones for polishing blades and creating the finest of bevels. Before use, saturate a Japanese stone in water for 20 minutes or more. Before sharpening, build up a slurry on the surface with a small chalk-like Nagura stone. Treat these stones with care because they will break if dropped.

More expensive are diamond stones, with particles bonded to a plastic or metal base. These are fast and hard-wearing, and cutting fluids make them effective although you can also lubricate with water. For touching up TCT router cutters, a thin diamond slipstone is very useful.

Both natural and synthetic oilstones are hard-wearing but limited in grades, and natural Arkansas stones are quite expensive. Traditional synthetic stones are made of either silicon carbide or aluminium oxide particles, each lubricated with thin oil. Combination oilstones have a different grit on each side, with the coarse surface used for occasional regrinding work.

For honing carving chisels and gouges, use a special slipstone. These include tapered and conical shapes to fit inside tool bevels.

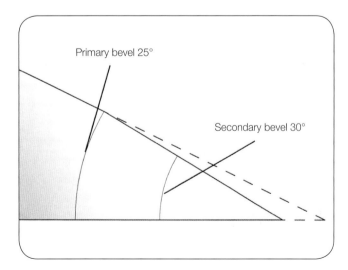

Sharpening an edge tool

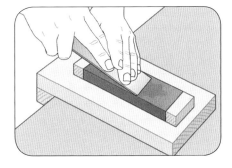

1 Apply the appropriate lubricant to the sharpening stone. If sharpening freehand, grip the blade in one hand and place the grinding bevel on the stone. Add pressure to the back with two fingers of your other hand. Raise a few degrees and move the tool forwards in a figure of eight pattern. You only need to hone the first 1mm (1⁄32in) of the cutting edge.

2 This process produces a wire burr, removed by laying the rear of the blade horizontally on the stone. Using two fingers to keep it completely flat, move the tool sideways once or twice. Make repeat strokes with the tool bevel down, then flat on its back again to finally clear the burr.

Regrinding

The more you sharpen a tool the wider the secondary bevel becomes, until the blade needs regrinding. For this it is best to use a powered whetstone grinder, which has a continuous flow of water to cool the tool; the stone can be vertical or horizontal and rotates slowly to avoid overheating. On a conventional high-speed bench grinder the steel may overheat, leading to a loss of temper and reduction in hardness of the cutting edge. High speed bench grinders are fine for woodturning tools, but they should be avoided for general woodworking tools.

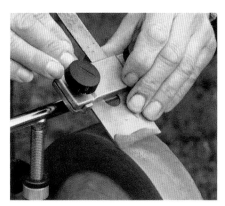

Honing guides

It is essential to keep a chisel or plane blade at a constant angle when honing. Until you get used to gripping the tool and maintaining this position, a honing guide is an important aid. You insert the blade and tighten a screw to present the edge at the correct angle to the stone, while a roller enables you to move the blade along the stone.

Sharpening saws

It is possible to sharpen your handsaws in the workshop, though it takes practice and skill to do a good job. Saw doctors (check your local tool store) have the equipment and expertise to sharpen bench saws much more accurately than even a skilled woodworker can. If you do want to try sharpening a saw yourself, obtain a cheap, second-hand tool to practice on first.

However, you can sharpen the very small saws. Dovetail saws are best sharpened using a 100mm (4in) precision saw file at right angles to the blade. The technique is to settle the file into the gullet of each tooth, then to give one stroke per tooth. Sharpen saws like this little and often, rather than waiting for them to get dull and useless.

files and rasps

For shaping timber (lumber) in any direction the simplest tool is often a file or rasp. Files have small, uniform teeth and produce a fine finish, while rasps have coarse teeth for cutting wood rapidly. You can also shape plastics or metals with a file. With these abrasive tools the cutting action is on the push stroke and both tools are produced in several patterns and sizes.

Files

Much less aggressive than a rasp, a file will eradicate most of the marks made by the coarser tool. Each face is covered with rows of teeth, unlike the individual teeth on a rasp. Apart from tooth pattern, files are classified by their finished cut. Smooth cut is the finest, bastard cut is coarsest and in between these two is the second cut. Occasionally you will need a file that can be used on metal – for example, the edges of a plane body may need softening or a screwdriver tip need reshaping. A good multipurpose tool for such work is a rectangular single-cut mill file. A set of needle files is useful for intricate shaping in timber (lumber) as well as metal or plastic, creating access to areas that would be tricky with larger tools. At about 150mm (6in) long, these are much smaller than standard files.

Related info

Scrapers and spokeshaves (see page 50)

Teeth patterns

For woodworking the most effective teeth patterns are double-cut and single-cut. Teeth run diagonally across the face of a double-cut file in both directions, which cuts fast but creates a rough surface to the timber (lumber). Teeth are also diagonal on a single-cut file but run in only one direction, creating a finer surface if you cut lightly.

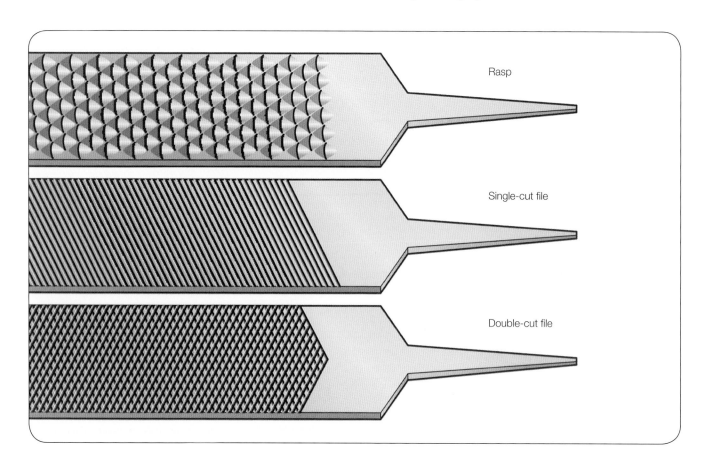

Rasp

Single-cut file

Double-cut file

Rasps and rifflers

For shaping timber (lumber) very rapidly or getting an uneven surface reasonably flat, rasps are aggressive tools that give a coarser cut. Available in several patterns, the most useful rasp has one face half-round for curved work, while the reverse face is flat. Teeth may either be formed by machine or cut by hand, which will produce a more random pattern and gives a cleaner finish to the timber (lumber); hand-cut tools cost more to buy. Bent into an S–shape rather than straight, the riffler has an arched face at each end. Some have coarse teeth like a rasp and tend to tear the wood rather than cut it. Although good for fast waste removal, use a finer rasp or file to remove marks before final tidying up with abrasives.

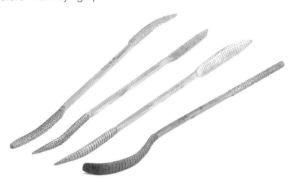

Similar to a plane with handles at both ends, the Surform tool uses steel blades with uniformly punched teeth, creating a coarse cut. Holes produced by the punching process mean waste material clears more easily, so there is less tendency for them to clog. You can use Surforms on all sorts of materials, including plastics and nonferrous metals, and their replaceable blades make them fairly cheap. Straight and curved formats are available.

Microplanes have a cutting action like a cheese grater; their extremely sharp teeth produce a finer cut. Blades are narrow and more delicate than a Surform, and may flex if used too roughly. Patterns include flat, curved and V-groove, all of which are interchangeable.

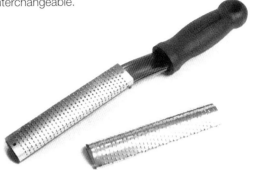

Adding handles

Files and rasps are not always supplied with handles, particularly better quality tools. For increased control and safety always fit handles to the tangs – they can either be bought or made in the workshop.

Keeping files and rasps clean

Use a wire brush to clean debris from rasps. A blowtorch will shift resin deposits, but use with great care. A file card, shown below, is designed to clean files by drawing it across the file face.

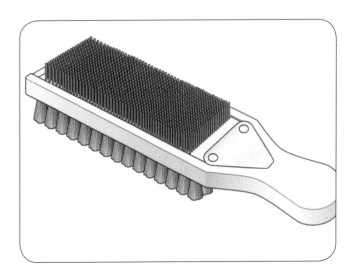

drills and bits

Accurate drilling, or boring, is a basic skill in the workshop, with the hand drill one of the simplest tools to use. Forerunner of the electric drill, it is easier to control than its power equivalents – particularly if using small diameter bits – and does not require a power supply. In Europe there tends to be a greater range of drill bits available in metric rather than imperial sizes.

Related info

Cordless power tools
(see page 62)
Electric drills (see page 66)
Pillar drill (drill press)
(see page 106)

Bradawls

The bradawl has a square-section blade and is the simplest tool for making holes. Instead of removing waste wood as with a drill bit, you twist the tip into the grain, forcing the wood fibres apart. This pilot hole enables you to insert a screw or accurately locate a larger drill bit.

Hand drills

A hand drill is operated by rotating its side handle, which is fixed to a gear wheel with teeth around the rim. These teeth mesh with either one or two pinions, causing the chuck to revolve. Bits are inserted in the three-jaw chuck, which has a capacity limited to about 8mm (⁵⁄₁₆in). Handles are usually plastic, or may be hardwood on older tools.

Braces

Less commonly used these days, the swing brace excels at drilling large diameter holes. The ratchet action means you do not have to fully rotate the handle, making the tool ideal for use in cramped conditions. You can swing the chuck clockwise or anticlockwise by swivelling a cam ring in either direction, with or without ratchet action. The four-jaw chuck accepts both square- and round-shanked bits, with a capacity of 12mm (½in). For driving or removing large screws you can fit special bits – the brace creates massive torque.

Choosing a bit

There is a daunting array of drill bits on the market, but not all will fit every type of drilling tool. Lip-and-spur (dowel) bits and twist bits, for example, can be used with hand or power tools and machinery, while other bits have specific shanks to suit particular chucks. For woodworking, lip-and-spur bits with their brad points are easy to position in wood. They cut particularly clean holes, but cramp (clamp) an off-cut underneath the workpiece for a clean exit.

If you need to drill various materials a good compromise is offered by engineer's twist bits, also called HSS bits. Designed for metal, these also perform well in wood and are perfect for drilling tiny holes. When drilling larger diameters, it is wise to mark the hole first on the workpiece with a centre punch.

Originally designed for swing braces, auger bits are a favourite for power tools but should never be used in a pillar drill (drill press) or vertical drill stand. As the bit revolves, the centre thread is drawn into the wood, with the outer spurs cutting the actual hole, while a spiral on the shank clears away waste material. Older Jennings pattern and auger bits tend to have square taper shanks, for secure gripping in the chuck.

For drilling holes that are not a standard diameter you can use an expansion bit with a brace, which has one spur cutter that you adjust outwards from the shank and lock with a screw. With this it is possible to drill holes up to 75mm (3in) in diameter.

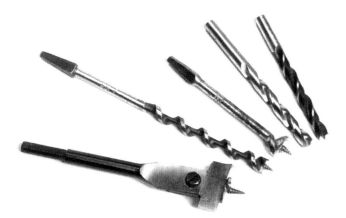

Drill bits, clockwise from bottom left:
1. *Expansion bit.*
2. *Auger bit.*
3. *Centre bit.*
4. *Twist bit.*
5. *Lip-and-spur (dowel) bit.*

Countersink bits

For neatness you will normally need to recess the head of a woodscrew flush with its surrounding surface. Once you have drilled the clearance hole for the screw, lightly rotate a steel countersink bit just in the top of the hole. To mirror the underside of the screw head a 90-degree bit is suitable, though you can also get 60-degree countersinks. Countersinks are either rose (fluted) or snail pattern, which give a smoother recess. You can slide special hollow countersinks on to twist bits, locking them with small hex screws.

Warning

Always hold timber (lumber) in a vice or cramp (clamp) it to the bench when drilling. Never hold the workpiece in your hand when using a drill of any sort.

hammers and mallets

Used for tasks such as tapping in a cabinetmaker's tiny veneer pin to driving in a hefty roofing nail, you will find at least a couple of hammers in every workshop. Frequently used to knock together joints, such as dovetails, or to make a quick routing jig, for such a simple tool there are many versions and sizes. Many hammers feature anti-vibration design to reduce fatigue and wrist injury, while state-of-the-art materials such as titanium and graphite make some futuristic tools particularly strong, yet light and comfortable to use.

Hammers

Unlike most woodwork tools, hammers are chosen not by size but by weight – a lightweight steel head will not drive in a 150mm (6in) nail, for example. Ash or hickory is commonly used for traditional hammer handles so, as well as being economical to buy, you can replace a damaged hardwood handle easily. The handle end is wedged tightly into the head's socket, and made secure by steel wedges driven into the handle from the top. Larger claw hammers often have rubber-sheathed fibreglass or steel handles, rather than hardwood.

The most useful general purpose tool is undoubtedly the claw hammer, a good weight being 450g (16oz). Use the claw to remove bent nails easily without causing damage, while the head is capable of striking most nails effectively. For the biggest nails the heavier 550g (20oz) size is better, but this may become tiring to use after a while.

For smaller nails the cross pein hammer is a better option; a useful weight would be around 225g (8oz). Use the wedge on the head to begin striking a nail while holding it steady on the wood. When the nail is firmly seated, use the opposite face of the hammer to drive it home. Even smaller is the pin hammer, which weighs about 100g (3½oz) and is designed for panel or veneer pins. Some woodworkers prefer a nylon or rubber-faced hammer for constructing carcasses or joints, because these are less likely to damage the wood.

Hammer grip

Always hold a hammer at the end of the handle so you can swing it effectively, with greater control.

Related info

Screws and nails
(see page 277)

Nail punches (nail sets)

Unless using fixings for general carpentry work, most interior woodwork that relies on nails looks bextter if the heads are concealed. To do this, drive the nail or pin below the wood surface with a steel punch (nail set); these are made with tips in several sizes. When all nails have been punched, you can then fill the holes with a matching filler.

Mallets

Although a heavy-duty chisel may occasionally have a steel end cap for striking with a hammer, it is always far better to use a wooden mallet with any edge tool (apart from Japanese chisels). It is safer because the mallet's striking area is bigger – you are also unlikely to dent or split the chisel handle. You can make a mallet in the workshop from any hardwood off-cuts, though commercially-made tools are normally beech.

screwdrivers

There should be at least a couple of screwdrivers in your toolkit, even if you rely on cordless tools – in fact, you will have greater control with a manual screwdriver than when using a powered version. To cope with any screw you are likely to come across, a decent screwdriver set should contain slotted, Pozidriv and Phillips head tools. Ergonomics are as important on a screwdriver as with other hand tools, and handle comfort is continually evolving.

Screwdriver design

Although there is a growing variety of screw patterns to choose from, most woodworkers use either cross-head or the more traditional slotted screws. Cross-head screws are favoured by cordless tool users because the screwdriver bit is less likely to slip when inserting them. When working with softwoods you also do not usually need to drill a clearance hole first, making this combination fast and convenient, although the final appearance is not always that neat. Slotted screws remain popular, however, especially on projects with brass hardware such as butt hinges, handles and catches.

Screwdriver handles are generally plastic apart from traditional slotted cabinet screwdrivers, which are oval, lacquered beech. Textured handle types provide a better grip than smooth PVC handles, which can slip and are not so comfortable over time. Screwdriver blades are usually made from chrome vanadium steel, with the tips often hardened to increase durability. Magnetic tips are helpful when driving small screws, while to improve screw grip some tools have sandblasted tips. Large, heavy screwdrivers can have square – rather than the regular circular – section blades, which allows you to attach a spanner (wrench) to increase torque when removing or inserting stubborn screws.

Crosses and slots

The most familiar cross-head screwdriver is the universal Pozidriv pattern, which has a cross projection at the tip of the blade designed to fit a recess on the head of a Pozi screw. Extra fins between each of the main cross projections on the tip improve grip, so there is a reduced chance of the tool slipping and spoiling the screw. The older Phillips pattern screwdriver is also still used, though this only has the single cross projection on the tip so is more susceptible to slipping. Don't be tempted to mix a Pozi screw with a Phillips tool, or vice versa. A slotted cabinet screwdriver has a flattened tip, either ground parallel or tapered to fit the screw slot – a flared tip is stronger for extra torque. When a screw

Related info
Cordless power tools (see page 62)
Screws and nails (see page 277)

must be recessed below the surface, however, use a screwdriver with a parallel blade because a flared tip is likely to bind in the hole. A wide, flat section of the blade beneath the handle allows you to use a wrench for extra torque. Made for screws from No.6 up to No.14, you can obtain a slotted cabinet screwdriver with blade lengths up to about 250mm (10in).

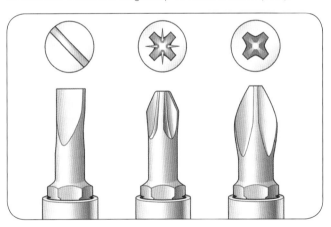

Slot head, Pozidriv and Phillips heads.

Other screw patterns

One of the most common new screw patterns is the Torx screw, which requires a special screwdriver. If you have machinery a set of hex keys will also be useful – hex screws are used widely in engineering and are also frequently found on woodworking machines.

power tools

Most woodworkers probably have at least one power tool, even if they are hand tool devotees, because power tools remove much of the hard work in woodworking. Planing heavy boards to thickness or sawing sheet materials can be tedious, but power tools speed the process, freeing time to hone important hand tool skills. Cordless tools are becoming more compact and innovations in engineering design and electronics mean performance often matches their mains-powered cousins. Although safety is a top priority, noise, vibration and ergonomics are crucial aspects of power tool design. Most modern power tools offer excellent value, but professional equipment will give greater precision, reliability and increased lifespan.

cordless power tools

For convenience a cordless power tool is unbeatable, whether you are drilling holes, driving in screws or sanding. A cordless saw can be particularly handy when buying wood at the timber (lumber) yard – with no mains electricity supply needed you can use them anywhere, although it pays to have a second, fully charged battery with you. Innovations in battery technology mean even budget tools can be as powerful as their mains-powered rivals. Fast chargers make it easy to recharge a battery while taking a break, although you will need a power supply to do this. With no trailing cable, cordless tools are also safer.

Drill/drivers

Arguably the most popular cordless tool, the drill should be part of any basic toolkit. If it is fitted with a collar that you can rotate at the front it is a drill/driver, designed to insert or remove screws as well as drill holes. The collar is the torque control or clutch, typically with 16 or more settings, which enables a screw to be driven flush with the wood and prevents you stripping the screw head. When working in tight spaces (such as fitting cupboards) a compact 10.8V or 12V drill may be all you need. For most woodworking tasks you will find a 12V or 14.4V drill suitable, since it has enough power to drive most screws. A more powerful 18V tool has greater torque for those larger diameter holes and screws, but will be heavier.

If the tool has two – or even three – variable speeds, you can select the speed via a lever or button on top of the body. Choose a slow speed range, which produces high torque, for driving screws or drilling large holes; for drilling smaller holes select a high speed with lower torque. When driving screws into wood select a clockwise direction, reversing the spindle direction to remove them – the forward/reverse selector button is usually close to the trigger. A built-in worklight is a handy feature when lighting conditions are less than ideal. On smaller tools equipped with keyless chucks bit capacity may be 10mm (⅜in), although 13mm (½in) is more common. You can tighten single-sleeve chucks single-handed, although double-sleeve chucks require two hands.

For greater versatility a combination (combi) drill is a good option – besides the standard rotary action, this tool also has a hammer action for use on masonry. A detachable side handle prevents the tool catching and twisting, a frequent problem when drilling masonry.

Related info

Electric drills (see page 66)
Portable saws (see page 68)
Portable sanders (see page 72)
Power planers (see page 76)
Dust control (see page 20)
Safer woodwork (see page 22)

Screwdriver bits

Hexagonal-end screwdriver bits fit into specific (¼in) chucks and include slotted, Pozidriv, Phillips, Torx and hex patterns. You can insert longer bits directly into a drill/driver or impact tool, although short bits are easier to use with a quick-release holder, so swapping bits is fast. You can also buy drill bits with hex ends.

Impact drivers

An impact driver is built for driving screws rather than drilling, so it offers far more torque (typically 95Nm) than a normal cordless drill. The most powerful drivers can insert heavy screws up to M16 in size, with typical no-load speeds up to 2800rpm. Although standard drill bits will not fit the hex pattern ¼in chuck, you can buy adaptors.

Circular saws

Cordless saws, with power ratings up to 36V and speed around 4000rpm, are perfect for cutting sheet materials to size. With a TCT blade set at 90 degrees, cutting capacity ranges from 22mm (¾in) on compact trim saws up to 54mm (2⅛in) on larger tools with blade diameters of 86–165mm (⅜–6½in). For bevel cutting you can tilt the base plate to 45 degrees. Circular saws drain batteries quickly, although thin-kerf blades increase the time between charging. Spindle locking makes blade changing a simple task, and a parallel guide fence is standard.

Jigsaws

Like their mains-powered equivalents, most cordless jigsaws have rapid blade change and you can also tilt the base plate to 45 degrees for bevel cutting. When set at 90 degrees, depth capacity can be as much as 135mm (5¼in) on heavy-duty models, with a stroke rate per minute of up to 3000spm. Variable speed is a very useful feature to have and pendulum action is virtually essential.

Planers

From shooting door edges to adding a chamfer on a piece of furniture, the cordless planer is a useful tool outside the workshop. Its cutterblock normally contains two disposable carbide knives, rotating at 13,000rpm. These give a nominal planing width of 82mm (3¼in) and you can adjust rebating (rabbeting) depth to about 15mm (¹⁹⁄₃₂in).

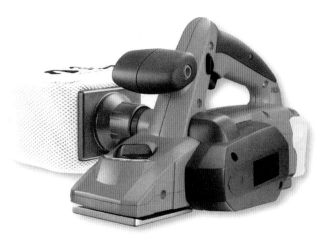

Sanders

Compared with mains-powered tools cordless sanders tend to be more compact, while bases are often delta-shaped and are fitted with hook and loop abrasives. You will find this tool handy for sanding cabinets, drawers and finishing furniture, although not all models have a dust bag or collector box.

Compact screwdrivers

For mounting cabinet fittings and hinges a small pistol-grip screwdriver may be useful, These are powered by a built-in battery (from 2.4–4.8V), but although they are cheap the charge time can be several hours. The handle may swivel to suit your grip and more elaborate models will have a torque collar and forward/reverse drive direction as standard, As with impact drivers bits are slotted into a ¼in hex chuck – the set of bits may be stored inside the body itself.

Cordless technology

Without a charged battery a cordless tool will not function. Voltages and capacities may be standard but the batteries themselves are not, so you cannot expect to fit a pack from one manufacturer into a tool of another brand. Each battery is made up of several 1.2V cells added together, so voltages are multiples of this (10.8, 12, 14,4, 18V, for instance). Battery capacity is measured in amp hours (Ah), which gives some idea how long the battery will power a tool before needing a recharge. To indicate remaining capacity, some batteries feature a digital or visual display.

Cordless technology has evolved rapidly over the past few decades: 7.2V tools were once a popular choice for many woodworkers and anything rated at 9.6V was considered a serious professional tool. Creeping upward, 14.4V and above now tend to be the preferred voltages, while in the construction industry many woodworkers choose 18V tools, which give a

Buying cordless tools

It makes sense to buy cordless tools of the same brand and voltage each time. Check the batteries are interchangeable, and then you can buy further tools bare (without the battery or charger) to economize in future.

good balance between performance, reliability and weight. Should you need even more power, some tools are available in 24V or even 36V – although this professional equipment will have a price tag to match. At the other end of the scale, compact tools with 10.8V or 12V batteries are very handy for working in tight situations. Although most cordless tools are provided with at least two batteries some manufacturers offer a bare (without battery or charger) option, which makes it more economical to add tools of the same brand with an identical battery fitting.

Some fast battery chargers will take just 15 minutes, though typical recharge time is up to one hour. Although they are far more expensive, sophisticated diagnostic chargers will analyze battery condition and history, and then charge appropriately. These days you can generally leave the battery in the charger without damaging it, because most units switch from fast to trickle charge rate when the battery is full. Some units will charge both NiCd and NiMH batteries. Although there are units that can be powered from a car, most chargers must be plugged into a mains supply.

Warning

For safety, always remove the battery before changing the blade or knives on a cordless saw or planer.

Battery basics

They may be cheaper, but nickel cadmium (NiCd) batteries are more harmful to the environment than other rechargeable types and must be recycled correctly because cadmium is a toxic metal. Another problem with this type of battery has always been the "memory effect" – the battery had to be fully discharged each time before recharging, otherwise its capacity would wane, so reducing remaining battery life. In effect, the battery would eventually not operate even at maximum capacity. If the tool was not used for some time the NiCd batteries would also discharge themselves, so would need recharging again before use. Most NiCd batteries don't reach optimum performance until they have gone through several charge and discharge cycles. However, NiCd batteries need less complex chargers and will operate at lower temperatures – although they are less powerful at up to 2.4Ah. With amp hour ratings up to 3.5Ah, nickel metal hydride (NiMH) power packs are now preferred by most power tool brands.

By comparison, a lithium ion (Li-ion) battery does not need to be fully recharged every time. Although more expensive, the other advantages of Li-ion are that they are lighter, suffer no memory effect and have little negative discharge. In terms of performance, a 28V Li-ion power pack will give up to twice the operating time of an 18V NiCd battery.

electric drills

Cordless tools may be more popular, but the classic mains-powered drill is a reliable workhorse for both workshop and general maintenance jobs. Unlike a cordless drill it will never run out of power in the middle of a job, although you are restricted by cable length. For increased accuracy when drilling holes you can mount the tool in a vertical stand, so it becomes a portable bench (pedestral) drill. For dense masonry, an electric drill with hammer action is far more efficient.

Power drills

At the core of a drill is its motor, which will be about 250W on compact tools and up to 1200W on heavy-duty, percussion models. Increased power is better for masonry but a bigger motor increases weight, making the tool tiring to grip if drilling into wood for more than a few minutes. Power drills are equipped with one or two speeds, and you select gears by a lever or button on the motor housing. Two-speed drills are more versatile, typically being rated 0–1000 and 3000rpm. Low speed means high torque, best for drilling large diameter holes or driving screws. High speed means lower torque, best for smaller holes. Variable speed is a handy feature, adjusted via a dial on the on/off trigger or casing. A lock-on button will keep the motor running without keeping your finger on the trigger. When driving in or removing screws you can alter chuck direction with a forward/reverse lever near the trigger, although there is no torque collar so you cannot adjust screw depth like you can with a cordless screwdriver. Larger tools often have a detachable side handle for better drilling control, which may include a sliding rod that you can tighten to set a maximum depth. For wood, metal or plastic the power drill works with a simple rotary action. Impact or percussion drills also have hammer action for masonry and you can swap between these two functions with a lever or button. Only use TCT bits for masonry work.

Related info

Drills and bits (see page 56)
Cordless power tools (see page 62)
Pillar drills (drill presses) (see page 106)
Dust control (see page 20)
Safer woodwork (see page 22)

SDS hammer drills

Built for use on hard masonry and concrete, an SDS hammer drill uses unique bits that lock quickly but securely into a special chuck – you cannot use these chucks with conventional drill bits. SDS bits have a standard shank that fits into the quick-release chuck, although on most SDS tools you can swap this for a normal keyless chuck so you can use ordinary drill bits.

Chuck choice

These days most chucks are keyless, with capacities of 10–16mm (⅜–⅝in) – the most common size is 13mm (½in). Some chucks feature automatic spindle locking, so you can tighten them with one hand, while others require both hands. Older drills tend to be fitted with keyed chucks; to prevent the key being lost, tape it to the cable.

Using a drill

For safety, always unplug the drill before inserting or removing a bit. Make sure the wood being drilled is either cramped (clamped) to the bench or held in a vice. Start slowly by gently squeezing the trigger and try to keep the drill perfectly square to the surface, whether drilling vertically or horizontally. If you find lining up the drill difficult, ask someone to watch as you work.

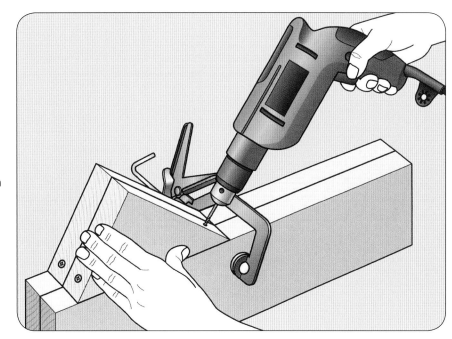

Using a drill stand

Most electric drills have a 43mm ($1\frac{11}{16}$in) neck, so you can cramp (clamp) the tool in a stand for vertical drilling. Holes drilled this way will be more accurate than if made freehand, and both hands are free to guide the work. Fit a piece of plywood or MDF on the stand base first to stop any possibility of going down into the base accidentally – this will also prevent the wood from breaking out underneath and makes it easy to attach a fence. A fence is very useful if you have several holes to make along an edge. For stability, cramp (clamp) or bolt the drill stand to a bench top. Never use an auger bit if you are using a drill stand – these have centre threads so the timber (lumber) will be lifted from the table by the drill, which could be dangerous.

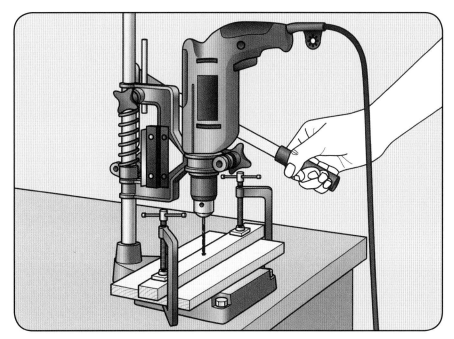

Using an RCD

Get into the habit of plugging a power tool into an RCD (residual current device). In an emergency this will shut off the current instantly, preventing any danger of electrocution.

portable saws

For cutting timber (lumber) and sheet materials portable saws are indispensable power tools, whether inside or outside the workshop – in fact, without other machinery they are almost essential. You may have to cut up sheet materials outdoors if your workshop is small. A circular saw is the fastest way to make straight cuts, while a jigsaw is best for curved cutting. Most saws will cut material at least 50mm (2in) thick and should cope with hardwoods and softwoods equally well.

Related info
Saws (see page 40)
Cordless power tools
(see page 62)
Workcentres (see page 89)
Woodworking machines
(see pages 90–119)
Dust control (see page 20)
Safer woodwork
(see page 22)

Circular saws

A circular saw is used for cutting timber (lumber) to length, ripping boards to width and converting sheet materials to size. Cutting accuracy depends on build quality and fitting the correct blade, which should always be sharp. For narrow ripping you can slide a side fence into the cast alloy or steel base plate, cramping (clamping) it at the correct width. For bevel cutting you can tilt the base plate anywhere from 0 to 45 degrees, locking it tightly in place with a thumbscrew or lever. To adjust the blade cutting depth raise or lower the base plate, which pivots on the motor housing. Brush motors on circular saws vary from about 500W on basic models to more than 2000W on heavier, professional machines. Saw blades rotate at a fixed speed of about 5000rpm. To prevent the saw being accidentally started up, most tools have a lock-off safety button that must be depressed before squeezing the on/off trigger.

Circular saw blades

Blades on circular saws tend to be TCT and will cut both solid timber (lumber) and sheet materials. A combination blade is suitable for both ripping and crosscutting, with 24–40 teeth ideal for general woodwork. Diameters range from 165 to 355mm (6½–14in), with a cutting capacity of about 51–130mm (2–5⅛in) when the base plate is set at 90 degrees. To prevent the wood binding against the blade during a cut a riving knife is sometimes fitted behind it, although not all saws include this safety feature. For changing the blade, most saws feature a simple spindle lock system. Use the wrench provided to release the blade retaining nut – a lock button stops the motor spindle turning during this process.

Using a circular saw

Never remove the riving knife if there is one already fitted to the saw. The blade must pass right through the timber (lumber) or board, so adjust the depth so that a full tooth protrudes below the wood thickness for maximum cutting efficiency. If the timber (lumber) is too thick for the saw don't be tempted to cut from both sides, which could prove dangerous. Use a jigsaw fitted with a longer blade instead, or cut the material with a handsaw. You can mount a circular saw upside down in a workcentre, effectively turning it into a small table saw. This leaves both hands free to guide the timber (lumber) or board, which can be fed against a rip fence or placed on a sliding guide for crosscutting.

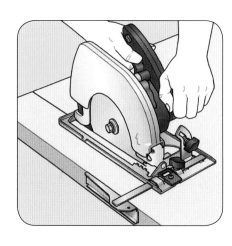

Guide rails

Some jigsaws and circular saws can be used with an aluminium guide rail. The base plate slides smoothly along a grooved track, which contains plastic strips, and rails can be joined together to form a longer fence. This is an ideal system for cutting sheet materials, since the neoprene strips underneath reduce slippage and minimize splintering.

Jigsaws

Use a jigsaw to cut wide, sweeping curves or fairly tight radii – it is fitted with a disposable, narrow blade and it is the width of this that determines the radius you can cut. When sawing a straight line with a jigsaw you will need to tidy up the cut afterwards unless you are only converting rough timber (lumber). On most tools the blade operates with an orbital movement, which is more efficient than a straight up and down action because it clears sawdust away and minimizes blade wear. Also called pendulum action, the degree of orbit is altered via a lever above the blade mechanism.

Jigsaws come in two body patterns – top (enclosed or D) handle and body grip – and some manufacturers sell identical models in both options: which you choose is down to personal preference. Motors are rated from about 350W on budget saws up to about 750W on heavier models. Variable speed is a useful feature, since it will enable you to cut metals and plastics as well as wood – it is controlled via a dial or varying finger pressure on the on/off trigger. Typical speed range is 800–3000spm and a lock-on button saves you having to squeeze the trigger constantly while cutting.

All but the most basic jigsaws feature rapid blade change: push a spring-loaded lever with your thumb, insert the blade and you are ready to cut. Older tools may use either a hex key to tighten the blade, a screwdriver inserted in the top handle or a locking knob on the handle itself. A steel roller supports the back of the blade as you cut– professional tools are fitted with guides to prevent sideways movement. When set at 90 degrees, cutting capacity is around 60mm (2⅜in) on basic saws, extending up to 135mm (5⁵⁄₁₆in) on professional models, while stroke length on both is about 25mm (1in).

For bevel cutting up to 45 degrees you can tilt the base plate either way, locking it with a lever or hex key. A plastic shoe clips over the base plate on some saws to prevent scratching on delicate surfaces as you cut. A plastic anti-splinter shoe slotted into the base plate stops chipping on plastic laminated surfaces. To keep the cutting line free of sawdust a built-in blower option is handy, while most jigsaws have a dust extractor outlet.

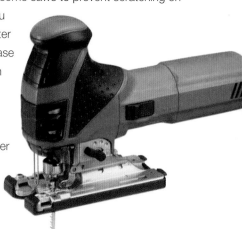

Jigsaw blades

Most jigsaws are fitted with bayonet blades; older tools use a universal format, tightened with a hex key. There is a wide choice of teeth patterns to match different materials, with lengths from 51–104mm (2–4in). Special reverse-teeth blades will cut laminate-faced worktops without splintering; use tungsten carbide grit blades on ceramic tiles, and fit appropriate blades for mild steel, alloys and nonferrous metals.

Using a jigsaw

Always cramp (clamp) the timber (lumber) down to a suitable bench or work surface before making a cut. Select the correct blade for the material and adjust the orbital action of the saw, if necessary. The greater the orbit the faster the cut, although this will result in a slightly rougher finish to the work. For a clean finish switch the orbit action to zero. Adjust the speed to suit the material; select a fast speed when cutting softwood and reduce speed if sawing thicker, dense hardwood or metals.

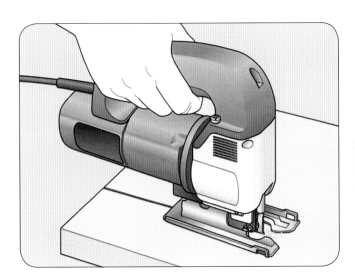

Reciprocating saws

Although more likely to be used for construction and house renovation work than in the workshop, the reciprocating or sabre saw is a specialized cutting tool for timber (lumber), sheet materials, metals, plasterboard and plastic. Ideal when sawing up recycled timber (lumber), it will even cut through embedded nails when fitted with the correct blade. Unlike a jigsaw, a reciprocating saw is held with both hands and its advantage over other power tools is cutting capacity. Fitted with a long, straight blade, it is possible to cut timber (lumber) up to 300mm (12in) in thickness. Rather like using a giant jigsaw, there is no fence for guiding straight cuts although curved cuts are easy to make. The motor is about 1000W, with variable speed of 0–3000spm. Cordless saws tend to have a smaller cutting capacity.

Checking the setting

Don't assume that the notched, centre position on the base plate of a jigsaw is set precisely at 90 degrees. Even on professional tools this setting can deviate as you tighten the locking mechanism, so the blade does not cut exactly at 90 degrees. Make a cut in thick timber (lumber) or sheet material and check the sawn edge is truly square. Readjust the base plate if necessary.

portable sanders

Many woodworkers have several power sanders in their workshops to suit different situations. Some tools will reduce stock aggressively with coarse abrasives or remove an old finish, while other sanders are designed to achieve a fine finish so no further hand sanding is necessary. Various abrasive grades can be fitted to these tools, although sizes and patterns are not always standard. Get into the habit of wearing a dust mask when using a sander, even though they generally have a dust bag or collector box. Most tools also have a dust outlet to connect a vacuum extractor hose.

Belt sanders

If you want to use recycled timber (lumber) – or simply sand large areas rapidly – the aggressive action of the belt sander is ideal. Its continuous abrasive belt rotates around two rollers, one of which is driven by the motor. Rated from 180W on budget tools to 1200W on professional sanders, the motor is either mounted in-line or transversely over the belt. On more sophisticated sanders you can select variable speed via a thumbwheel; typical speeds range from 250–450m meters (800–1500 feet) per minute. To remove the belt open out a side lever, releasing tension on one of the rollers – the new belt tensions automatically when you flip the lever back again. Once fitted, adjust belt tracking across the rollers to keep it running in line by rotating a knob, if fitted – some sanders adjust tracking automatically. To maintain central tracking, rollers are usually slightly cambered. Sander belt widths are 76 and 104mm (3 and 4in) – wider machines are better for wider surfaces although heavier to use. For static sanding, some sanders can be fixed upside down on the bench top.

Related info
Cordless power tools (see page 62)
Sanders (see page 112)
The sanding process (see page 244)
Dust control (see page 20)
Safer woodwork (see page 22)
Surface preparation (see page 242)

Sanding skills

The abrasive on a belt sander will clog sooner or later. You can unclog it and make it last longer by holding a rubber cleaner against the belt as it runs.

When sanding take care not to hold the tool stationary or you will rapidly sand a hollow in the surface of the wood.

Orbital sanders

The orbital sander is fitted with a vertical motor, with an off-centre shaft that rotates in an eccentric orbit. The platen (base plate) is flexibly attached to the tool's housing, with any vibration reduced by a balancing counterweight. The abrasive sheet is attached to a rigid rubber or polyurethane backing pad, which is glued to the platen. Motors are rated from 150 to 350W, and operating speeds vary from 6000 up to about 22,0000pm. A lock-on button next to the trigger can be used to keep the tool running for continuous sanding. Ideal for fast stock removal, roughing sanders have an orbit diameter of up to 5mm ($\frac{3}{16}$in), but for a finer surface finish choose a finishing sander, which has a smaller orbit diameter of 1.5–3mm ($\frac{1}{16}$–$\frac{1}{8}$in). Most orbital sanders have a tendency to cause small swirls over the surface of the timber (lumber), which may not be obvious until a clear finish such as varnish has been applied. Remove these marks with a final sanding by hand, or use a random orbit tool with a finer abrasive.

Orbital sanders are made in two common sizes that relate to the abrasive paper needed to fit the rectangular platen. A full sheet of abrasive paper measures 280 x 230mm (11 x 9in), so half-sheet sanders will use paper measuring 280 x 115mm (11 x 4$\frac{1}{2}$in), while third-sheet sanders use paper 230 x 93mm (9 x 3$\frac{21}{32}$in). The actual pads are smaller than these sizes, so the ends of non-hook-and-loop paper can be tucked under steel retaining cramps (clamps). Some sanders feature cramps (clamps) plus a hook-and-loop (Velcro®) system. Holes in the platen enable dust to be sucked up through the tool via a rotating fan; debris is directed to an outlet, which may be connected to a dust bag or extractor hose. There must be matching holes in the abrasive paper for the extraction system to work effectively. Instead of dust bags, some sanders have collector boxes with reusable filters.

Random orbit sanders

A problem common with an orbital sander is that it leaves behind swirls on the surface; the random orbit sander is a much better alternative, because its off-centre pad creates a random motion that leaves fewer swirls. As well as rotating the pad moves eccentrically, so previous scratches are cancelled out by newer rotations. Disc diameters are available at 115mm (4½in), 125mm (5in) or 150mm (6in) and are formatted to take hook-and-loop abrasives. Power is rated from about 250 to 600W, with speeds of 4500–13,500opm. For sanding veneered work electronic variable speed is worth considering – reduce the rotation speed and there is less chance of sanding right through the veneer. For general purpose sanding work, some professional tools can be switched between random orbit and a more aggressive orbital oscillating action – select the mode by flicking a lever above the pad. Sanding discs are normally punched to allow for dust extraction, so make sure the holes are aligned with those in the pad when replacing the abrasive. You can also fit a polishing bonnet to a random orbit machine for waxing or buffing finishes.

Buying abrasives

If using a sander fitted with cramps (clamps) it is more economical to buy unbacked abrasives by the roll, cutting pieces to match your sander. You can buy rolls in several widths and lengths.

Palm sanders

The palm sander is a more compact orbital tool, which is normally gripped with one hand. The pad size is usually quarter sheet, measuring 115 x 100mm (4½ x 3¹⁵⁄₁₆in), and the abrasive paper may be cramped (clamped) to the pad or in hook-and-loop format. Similar to a finishing sander, orbit diameter is about 1.5mm (¹⁄₁₆in), while at around 200W motors are low-powered and offer a single speed of 14,000opm.

Paper punch plates

A paper punch plate is provided with some orbital and palm sanders, allowing you to fit plain, unperforated abrasive paper but still use the dust extraction holes in the tool's base. Once the plain paper is fitted to the tool it is pressed against the punch plate, and the spikes make the required holes. If no punch plate is supplied you can easily make a jig using pieces of dowel glued into a timber (lumber) off-cut – use a pencil sharpener to add points to the ends of the dowels.

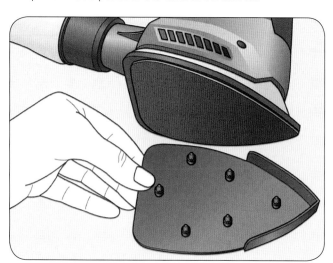

Hook-and-loop abrasives

Hook-and-loop abrasive papers may be convenient, but are more expensive than unbacked paper. It is much quicker to change from one grade of abrasive to another, with no paper cutting or cramping (clamping) to think about. Sheets are generally available pre-punched for both orbital and random orbit sanders.

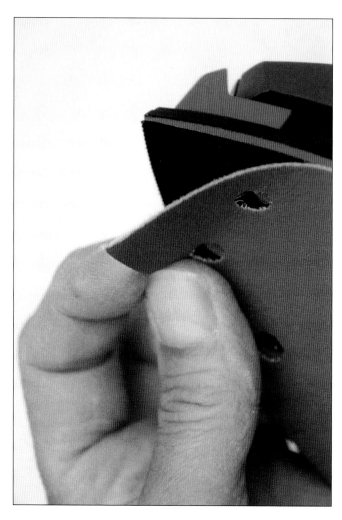

Detail sanders

The detail sander, with its triangular or delta-shaped sanding pad, is handy for reaching into the corners of cabinets, drawers and confined spaces. Abrasive paper will wear faster on the tip than the middle of the sheet – just rotate the backing pad when this happens or reposition the paper, because the fixing system is hook-and-loop. The motor on these tools is rated at around 200W, with variable speed up to 21,000opm. Detail sanders are versatile tools because you can fit most with polishing pads, cutting tools, scrapers, rasps and even saw blades, as shown below. Saw teeth may be exposed, but the oscillating action makes unguarded blades safe to cut thin material.

Multi-sanders

More DIY tools than anything else, multi-sanders are bigger than detail sanders and use interchangeable shaped abrasive pads, which can be used to sand flat, concave or convex surfaces. Delta-shaped hook-and-loop abrasives can be swapped for discs, while curved or V-pads can be added up front for sanding profiles. With around 150W of power, speeds are typically 11,000–24,000opm.

Using a sander

When preparing large areas, half-sheet orbital sanders are heavier and faster than third-sheet machines. When using an orbital, palm or random orbit sander, place the pad on the work before switching on the tool, but when using a belt sander switch on before placing it down on the work, keeping it moving while the belt is rotating.

Work through the grades

Always work down through the grades when using abrasives, starting with the coarsest grit. If you need to remove a visible scratch you may need to go back to the previous grit used. Clear varnish or lacquer will highlight the smallest scratch, so do not be tempted to cut corners during the sanding process.

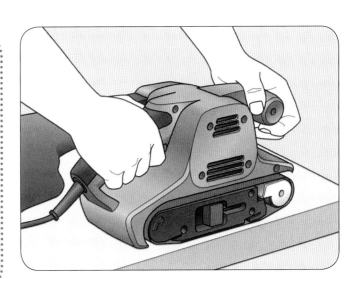

power planers

A power planer is a fast way to plane relatively narrow timber (lumber) or the edges of sheet materials. Ideal for trimming long boards, doors and panels, most tools are light enough to use vertically as well as horizontally. You can also cut 45-degree chamfers along an edge.

Power planers

A power planer allows you to take the tool to the timber (lumber), unlike a stationary surface planer (jointer) where material is taken to the machine and passed across rotating cutters. Rotating the front handle adjusts the cutting depth, raising or lowering the alloy base relative to the cutterblock knives. Before activating the on/off trigger the lock-off button must be depressed with your thumb. The maximum depth of wood you can remove in one pass varies between about 1.5 and 4mm ($^1/_{16}$–$^5/_{32}$in). Common planer width is 82mm ($3^7/_{32}$in), although there are heavy industrial tools with a width capacity of up to 312mm ($12^1/_4$in). The knives are carbide and are usually reversible and then disposable, so once both edges are blunt they are thrown away. Most planers are provided with a side fence to enable you to cut rebates (rabbets), to a depth of 24mm ($^{15}/_{16}$in) on professional models. To contain the chips the majority of planers are supplied with a detachable collection box or bag, but if you have a lot of planing to do it is more efficient to connect a vacuum extractor because on-board collectors fill quickly. Some tools have a built-in deflector that can be adjusted to eject chips from either the left or right. Typical power rating is between 650 and 850W, giving cutterblock speeds of up to 20,000rpm.

Related info
Planes (see page 44)
Cordless power tools (see page 62)
Planers and thicknessers (see page 108)
Dust control (see page 20)
Safer woodwork (see page 22)

Using a planer

Set the cutting depth to the minimum, then hold the planer with both hands and place the toe on the end of the timber (lumber) before switching on. Squeeze the trigger and guide the tool along the work, lifting it up when you get to the end. Repeat the stroke if necessary. Let the cutters reach a standstill before placing the tool down on a surface. For safety, many planers have a hinged plastic shoe at the back of the base plate, which flips down at the end of a cut. Keep the cable clear of the working area by placing it over your shoulder.

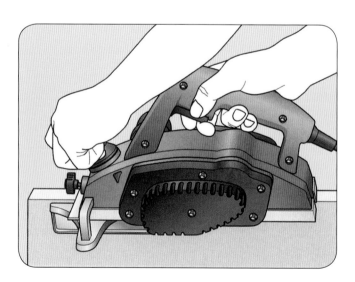

biscuit jointers

For making fast, simple butt joints in both solid wood and boards the biscuit jointer is the ideal tool. A curved slot is cut in the edge of one piece by plunging a rotating blade into the material, and then a matching slot is cut in the adjacent piece. A biscuit is glued in one slot, then both components are cramped (clamped) together and the joint is complete. You can make horizontal, vertical or angled joints up to 90 degrees, simply by moving the front fence – although it is essential that mating edges are planed straight and square for a tight joint.

Biscuit jointers

The biscuit jointer has a horizontal body, motor and drive mechanism similar to an angle grinder, and a rotary blade with a diameter of 100mm ($3^{15}/_{16}$in). This blade normally has six TCT teeth and rotates at around 10,000rpm, cutting a slot 4mm ($5/_{32}$in) wide. Teeth are only exposed when you push the tool forwards into the timber (lumber); the blade automatically retracts when you release pressure at the end of the cut, and spring-loaded plunge action means it stays concealed inside an alloy casing even when the tool is activated. Typically there are six preset depth settings, matching standard biscuit sizes – rotating the dial automatically sets the slot depth to suit the biscuit. Motor rating is about 500–700W, and there is a lock-off trigger on the body.

> **Related info**
> Biscuit jointing (see page 196)
> Rub joints (see page 182)
> Butt joints (see page 184)
> Dust control (see page 20)
> Safer woodwork (see page 22)

Fence setting

To enable you to line up the tool with pencil marks on the timber (lumber), the adjustable front fence has reference marks. You can tilt this for cutting slots at any angle from 0 to 90 degrees, with indexing at 45 degrees so mitred edges are as simple to joint as square edges. You can also raise or lower the fence so the slot can be positioned to suit different thicknesses of timber (lumber) – on thin boards it is important to cut the slot centrally in the edge. A dust bag is provided with most jointers, though you can usually attach an extractor hose instead.

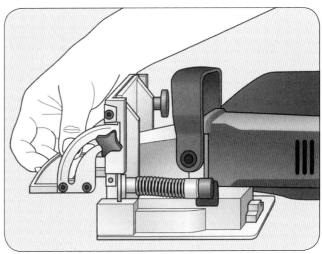

Changing the blade

Access to the blade is by removing the base plate on the tool, either by using a screwdriver or undoing a thumbscrew that is fast to release. A wrench supplied with the tool is normally used to remove the blade, in conjunction with a spindle lock button.

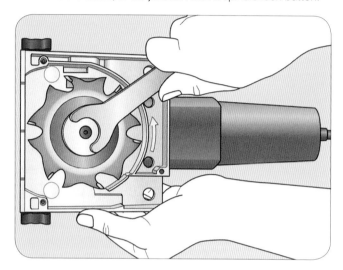

Biscuit assortment

There are three popular sizes of biscuit produced for different thicknesses of material –larger 86 and smaller H9 biscuits are less common. The No.0 is for timber (lumber) 6–12mm (¼–½in) thick; the No,10 is for timber (lumber) 13–18mm (½–¾in) thick; and the No.20 is for timber (lumber) 19mm (¾in) thick and above. All biscuits are the same thickness and fit a 4mm wide slot.

No.0 biscuit.

No.10 biscuit.

No.20 biscuit.

Other jointing systems

Recent innovations in jointing include a portable morticing system that uses loose, oval-shaped hardwood tenons. Instead of a circular blade, an elongated slot is cut with an oscillating drill bit. Unlike universal biscuits, which can be used with any make of biscuit jointer, this system is specific to one manufacturer. Precision power tool dowel joint cutters are also available.

Biscuit storage

To keep moisture out it is best to store biscuits in an airtight jar – they can become too tight for the standard slot if allowed to swell up.

routers and cutters

Without doubt, the revolutionary electric router is the most flexible and indispensable power tool in any workshop. It is capable of a wide range of cutting and shaping tasks, and you can fit it with a bewildering array of shaped cutters. With a little practice, even the unskilled beginner can shape edges, rout mouldings, cut housings, and make other joints quickly and accurately – previously you would have needed a box full of hand tools to carry out the same woodworking tasks. You can operate a router freehand, attach a side fence for making straight cuts or fit special cutters for curved edges. Mount it upside down in a table and it becomes a small-scale machining centre for working with both solid timber (lumber) and sheet materials. Workshop-made as well as off-the-shelf jigs increase the potential even further – in fact, the router is only limited by the creativity of its user, making this power tool unique.

Related info
Router jigs (see page 87)
Workcentres (see page 89)
Dust control (see page 20)
Safer woodwork (see page 22)

Router basics

The motor of the plunge router is suspended between two steel posts, and directly drives a rotating cutter at the end of its spindle. The posts are attached to a base plate, and enable the motor to move up and down on springs, while the cutter is inserted into a collet at the end of the spindle. Rotating at high speed, the cutter is lowered into the timber (lumber) by pressing down on the spring-loaded motor. For edge routing you can lock the cutter at a fixed point and feed it into the wood sideways, guiding the router by gripping the handles on each side. It is common for workshops to have two or three routers, so you can leave one tool set up for a specific task. As a first purchase a small, lightweight router with a ¼in collet is ideal to practise your routing skills while being able to control the tool safely. For bigger cuts in heavier material a more powerful heavy-duty router with a ½in collet is preferable – this will be suitable for making large mouldings, such as a raised and fielded panel for a cabinet door. Although more cumbersome than a smaller tool, soft-start electronics make modern versions of these routers easier to control than older models. A mid-range router is also an option if you want a compact but powerful tool that is easy to guide.

Small ¼in router.

Mid-range router.

Heavy-duty ½in router.

Power and speed

A small, ¼in router motor is likely to be about 750W, but this will increase to 2000W on a large, heavy-duty tool. Virtually all routers now offer electronic variable speed, which is adjusted by a dial found on the motor housing. Typically, the speed range is 8000–30,000rpm; select a low speed for large diameter cutters, and high speeds for small cutters. Sophisticated technology includes soft-start electronics, which will maintain a constant speed under load and provides electronic braking on professional routers. The power switch may be a basic slider button or a double-action trigger built into the handles – these make accidental start-up of the router unlikely, although this sort of switch can be tricky to use when the router is mounted upside down in a table.

Collets and cutters

The router cutter slides into the slightly tapered collet, which is attached to the threaded motor spindle with a nut. Once the cutter shank is inserted, tighten the nut with a wrench – you may need to use two wrenches for tightening on older routers. Deactivate the spindle lock mechanism; this stops the motor spindle rotating and is invariably a push button. Standard collet sizes are ¼in and ½in – ⅜in and metric 6mm, 8mm, 10mm and 12mm collets are less common. Cutter shanks – irrespective of brand – will match these sizes; larger routers are usually provided with two or three interchangeable collets.

Router bases

Always allow a cutter to reach full speed before it comes into contact with the wood, no matter how the router is being used. To do this safely, almost every router has a spring-loaded base so the cutter can be plunged down into the wood. Once at the depth selected you can lock it in position by twisting a side handle or flicking a lever, raising the cutter once the cut is completed. Router bases tend to be made of cast alloy and they may be D-shaped, circular or have two parallel edges. A hard plastic facing gives a smoother movement across timber (lumber) and helps prevent delicate surfaces, such as veneer, from getting scratched. You can normally screw the base into a table via threaded holes, or attach it to a variety of jigs. Some professional routers have a fixed base for greater precision when adjusting cutting depth. Although powerful, these tools can be easier to control because handles are closer to the work, lowering the centre of gravity. You do have to tilt the router to start the cut, and then lift it off again afterwards – a drawback for some routing tasks. However, a few heavy-duty routers are provided with both plunge and fixed bases, which are easily swapped over withdraw the motor unit from one collar, insert it into the second unit and lock together. This gives the advantage of having a fixed base router mounted upside down in a table continually, which you can remove easily for normal overhead routing tasks.

> ### Warning
>
> Never be tempted to mix metric and imperial sizes when it comes to router cutters. They may seem almost identical, but a 6mm collet is actually smaller than a ¼in collet and the same is true of 12mm and ½in collets. If you cannot slide a cutter shank into a collet easily, don't force it – and if it is too loose, you will not be able to tighten the cutter adequately so it will be unsafe.

Router with interchangeable fixed and plunge bases.

Adjusting depth

A portable router has a locking system for its plunge action, which sets the cutting depth. On a basic machine you tighten an adjustable metal rod with a thumbscrew; the rod then contacts a fixed stop on the base plate when you plunge the router down. However, most routers have a rotating depth turret, some with adjustable stops, on which you can set several precise cutting depths by rotating the turret as necessary. Built-in fine adjusters can be found on more expensive, professional routers – first set the cutter depth approximately, then fine-tune with a dial or knob. Plunge capacity is the greatest depth the motor unit and cutter can plunge – this may be up to 80mm (3⁵⁄₃₂in) on a large router or around 51mm (2in) on a smaller tool.

Three-position depth turret with adjustable stops.

Plunge depth rod and adjuster.

Fences

Every router is supplied with a guide fence for cutting parallel to a straight edge. Made from cast alloy or pressed steel, its steel rods pass through the router base and are locked in place with thumbscrews at the distance selected. A built-in fine adjuster wheel provides precise adjustment to the fence.

Fence with fine adjuster.

Dust extraction

A router ejects plenty of chips as it cuts, so expect the workshop to get covered if you do not have adequate extraction in place. Although fiddly to fit, some routers are equipped with clear plastic guards that clip into the base, limiting ejection of chips, while a dust outlet enables you to connect a vacuum extractor hose. More sophisticated routers may have a swivelling outlet on top of one of the plunge posts. Always wear eye protection as well as ear defenders when routing – a dust mask or respirator is also essential, particularly if working with MDF.

Router tables

When mounted upside down in a table, the router becomes an accurate small-scale machining centre. In this configuration the cutter will protrude above the table and you feed material across the surface, pushing against a fence or with the workpiece fixed to a template. Small tables often make use of the fence rods on the router to suspend it; on bigger tables the router may be cramped (clamped) with brackets or simply bolted underneath. The router's on/off switch must be easy to reach when using the tool inverted. On older models you may need to cramp (clamp) the power trigger to keep it activated while you are routing – for safety, plug the router into a separate NVR switch, ideally fitted below the table, so you can then leave the router switched on but still operate it independently. A channel set into the table that accepts a sliding mitre fence means that you can machine tenons and make other end grain cuts precisely. When routing with a template you normally remove the fence so the entire table surface can be used. Routers produce a lot of waste, so an outlet at the back of the fence is essential. Always use safety guards with a table, and also a pushstick when feeding timber (lumber) past a rotating cutter.

Palm routers

For edge work or when space is tight the compact palm router is a useful option. Unlike a full-size router, you grip this tool in one hand; it is perfect for shaping profiles, and cutting grooves for inlay or housings for hinges, as well as being easy to control. Often used without a fence, bearing-guided cutters run against a template or the edge of the timber (lumber). Quite powerful at about 600W, the palm router operates at about 33,000rpm and has a standard ¼in collet so it will accept a variety of cutters.

Router cutters

There is a vast range of cutters, or bits, for the router; they come in many shapes and formats, irrespective of collet size. You can fit profile cutters to create decorative mouldings along the edges of timber (lumber), while straight cutters are best for grooves, rebates (rabbets) or cutting joints such as tenons. Most router cutters are tungsten carbide tipped (TCT) – although high speed steel (HSS) cutters are sharper and produce a better finish on softer woods, TCT cutters hold their edge much longer. Always use TCT cutters on sheet materials, which can be very abrasive. Hone TCT cutters lightly with a small diamond stone, but for professional sharpening send them to a saw doctor.

Fit a bearing-guided cutter to a router and the cutter will follow either a straight or curved edge exactly. Usually secured with a hex key, the roller bearing may be either above or below the profile edge. Assuming that the edge has been cut straight and true in the first place, you will not need to use the guide fence on straight cuts.

Router cutters vary widely in price and this is reflected in the quality. A cheap set of mixed cutters may be fine for occasional routing work, although they are likely to be inferior to more expensive cutters on which quality will be higher. The expensive cutters will also remain sharp for longer, so economically it is a good idea to buy branded cutters individually as you need them. You should only use large diameter profile cutters (more than 51mm/2in) with the router inverted in a table.

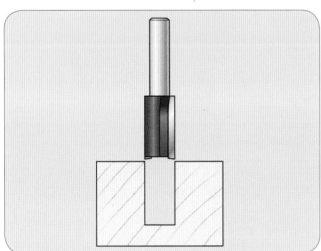

Straight cutter.

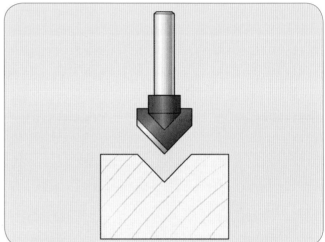

V-groove cutter.

Radius cutter.

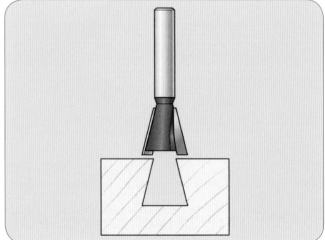

Dovetail cutter.

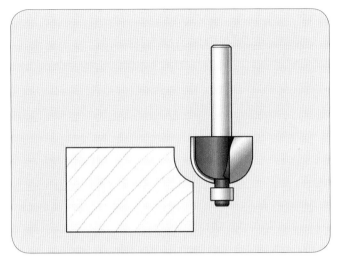

Cove cutter (bearing-guided).

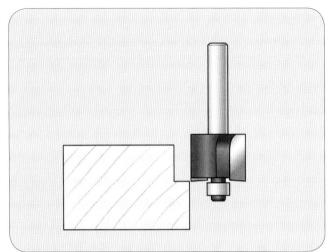

Rebate (rabbet) cutter (bearing-guided).

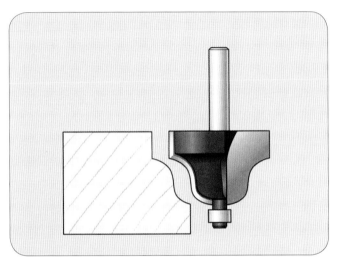

Roman ogee cutter (bearing-guided).

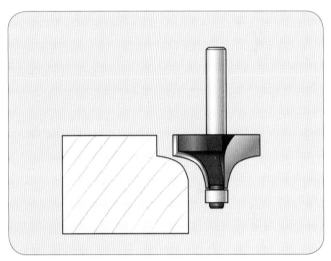

Rounding over cutter (bearing-guided).

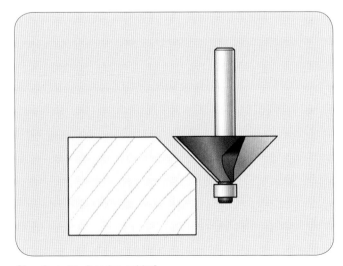

Chamfer cutter (bearing-guided).

Machining components

When machining several components to the same dimensions, always cut one or two spares so that you will not have to set up the machine again for a replacement if one gets damaged.

Direction of cut

When guiding a router always make sure you move it in the correct direction, particularly along edges. Should you feed it the wrong way the tool can jerk forward, making control difficult. Seen from above a router cutter rotates clockwise, so theoretically you should move the router anticlockwise around an outer edge, and clockwise around an inner edge. To some extent this will depend on the timber (lumber), however, because tear-out is possible –especially on wild grain. If tear-out does happen you may improve the quality of the cut by reversing cutting direction, although this requires extra care in controlling the router.

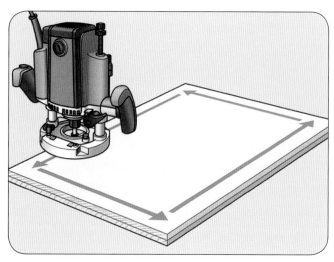

The direction of router travel is important on both internal and external edges.

Cutter care

Resin can build up easily on router cutters, especially if you are using softwood. Remove it by soaking the cutters in white spirit (mineral spirits) and rubbing gently with a fine brush.

router jigs

You can increase a router's potential by adding jigs and accessories, many of which are templates for making accurate joints in timber (lumber) and sheet materials. Although there may be a steep learning curve with some of these tools, once you have mastered a technique you will be able to carry out repeat joints fairly easily. Some joints demand considerable time and skill to cut neatly using hand tools, but with a router and appropriate jigs they can be made in a fraction of the time.

Related info

Mortise and tenon joints (see page 200)

Dovetail joints (see page 209)

General workshop safety (see page 92)

Dust control (see page 20)

Safer woodwork (see page 22)

Spare parts

Keep a couple of spare router mounting screws or bolts for the router in a safe place, because these can easily become lost.

Dovetail jigs

Using a dovetail jig usually involves cramping (clamping) a horizontal rail across the end of a board gripped in the vice. This rail comprises a row of fixed or adjustable fingers, and you position the dovetail template on top. To make the joint, fit a guide bush and cutter to the router, and then move it between the fingers. Cut both tails and pins with dovetail and matching straight cutters, adjusted to the depth required. Basic jigs may only cut lapped dovetails, while more elaborate versions will make through and sliding dovetails, mortise and tenon, and finger (box) joints. On the more expensive jigs you can alter the spacing, so the joint will appear to be hand-cut.

Mortise and tenon jigs

You can cut both mortise and matching tenon with this clever jig, using two different plates with a guide bush. Before routing, cramp (clamp) the wood in the vice at 90 degrees or at an angle to the cutter. This jig is perfect for making furniture, because you can also cut compound angles.

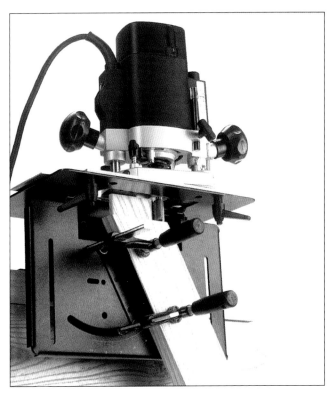

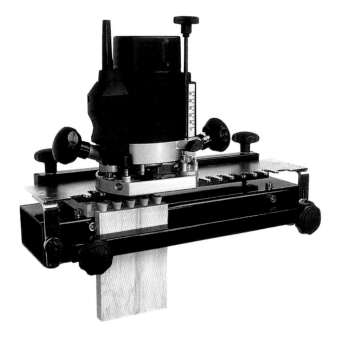

multi-tools

For small-scale woodwork – such as doll's houses, model making or musical instrument building – conventional power tools are often too large and cumbersome. The specialist techniques required fall within the scope of miniature power tools, which can be much easier to use and also less intimidating than their full-size cousins. Hand-held multi-tools are perfect for small-scale carving, routing, sanding and shaping work, and have a host of accessories for metalwork and plastics. Stand-alone machines offer limited sawing, planing and drilling capabilities and can be used safely on the bench top.

Hand-held tools

The basic multi-tool is similar to a hand-held miniature electric drill. You can fit a range of carving heads, router cutters, or sanding attachments into the chuck, secured via a spindle lock and wrench. Accessories have different spindle diameters, with collet sizes to match.

These tools are either mains-powered or may be cordless, so you can use them away from a power supply if necessary. Some mains tools are only rated at 12V – so you will need a transformer – while others plug directly into a 240V mains supply. Cordless multi-tools are supplied with a mains charger unit, taking about three hours to recharge. Batteries tend to be NiCd on cheaper models, although Li-ion is common on more serious tools. On mains mini tools motors are rated up to about 175W, and most offer variable speed that you adjust via a small dial. Speed range can vary from 5000rpm up to as much as 35,000rpm on professional models. You can mount these tools in stands for precision work, while router bases make them particularly versatile.

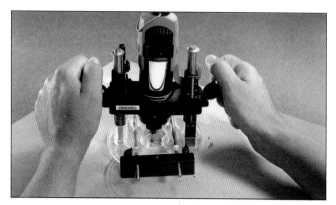

> **Related info**
> Cordless power tools (see page 62)
> Dust control (see page 20)
> Safer woodwork (see page 22)

Mini machines

It is possible to convert timber (lumber) on your kitchen table with mini machines. Table saws enable you to cut wood accurately to length and width at 90 and 45 degrees, although the maximum thickness is limited to about 22mm (⅞in). For curved work, bandsaws can cut material up to 80mm (3⅛in) thick. Planer thicknessers (jointer planers) offer a timber (lumber) capacity of about 80 x 40mm (3⅛ x 1½in).

workcentres

You may not need woodworking machinery to cut large boards or lengths of timber (lumber) accurately. An economical alternative is to fit certain power tools into a freestanding bench, or workcentre, so you can guide timber (lumber) above or beneath a router or saw with both hands, allowing greater control. When required, you can remove the tools again for hand-held operation.

Related info
General workshop safety (see page 92)
Universal machines (see page 118)
Dust control (see page 20)
Safer woodwork (see page 22)

Workcentres

Routers and circular saws are most frequently mounted in workcentres, although some units may accept jigsaws, sanders and planers. You normally cramp (clamp) the circular saw or router to a universal mounting plate, which then fits into the workcentre with the actual power tool above or below the wood, depending on the machining technique.

Workcentre routing

For overhead routing work you can mount a router on top of the workcentre. Either fit the tool into a sliding cradle so it travels across the timber (lumber), or cramp (clamp) it stationary so you can slide timber (lumber) underneath – it is possible to cut grooves and housings accurately with this set-up. A router becomes a small spindle moulder when inverted – insert a straight cutter and you can cut rebates (rabbets), grooves, tenons and shoulders with the fence adjusted. Change to a profile cutter and you will also be able to cut decorative edges. Cramp (clamp) pieces together when routing several identical components. If using a template, remove the fence and use bearing-guided cutters.

Workcentre sawing

Mount the circular saw above the workcentre, with the tool guided along parallel rails, to crosscut timber (lumber) or produce grooves. With the wood positioned on a lower height adjustable table you can cut housings and tenons, while stops allow you to saw from 45 to 90 degrees. With the saw fitted upside down the workcentre becomes a table saw; you can adjust the fence provided and rip timber (lumber) and sheet material. Crosscutting and angled sawing are accurate using a sliding mitre fence, while tilting the blade enables you to cut compound mitres.

Safety

When using the workcentre as a table saw, always fit the crown guard and riving knife.
If it is mounted overhead, check the blade will not foul the workcentre before you switch on the saw.
Have a pushstick ready when ripping timber (lumber).

woodworking machines

A single woodworking machine can cut our a great deal of time and drudgery, particularly with planing or sawing. Once wood has been prepared exactly to size, you can concentrate on finer hand tool techniques. Machines are freestanding units rather than portable, so the timber (lumber) is taken to the machine for cutting, shaping or sanding operations. Generally, the larger the machine the greater its timber (lumber) capacity and construction quality, while cast iron tables offer greater stability and accuracy so are a better longterm investment than alloy or steel. Safety and noise levels are major considerations, so try to see a particular machine in action before buying.

safety

Always respect the tool or machine you are using and avoid risks. Workshop accidents can often be avoided by being aware of safety issues and correct machining procedures. If a technique seems dangerous, there is usually an alternative method. Wear eye and ear protection when using routers, saws and planers especially, plus suitable face protection when sanding or woodturning.

General workshop safety

- If you don't know how to use a new machine safely, get professional advice or go on an appropriate course.
- Use a residual current device (RCD) with every
- mains power tool. If you should cut through the cable the RCD detects the change in current in milliseconds, breaking the circuit to prevent electric shock.
- Make sure you have enough mains sockets installed around the workshop to avoid trailing cables across the floor. Use a qualified electrician for any installation work.
- When changing a cutter or blade in a router or saw always unplug the tool first.
- Avoid using mains power tools outdoors in damp conditions – it is far safer to use cordless equipment.
- If you are feeling tired or have taken alcohol don't operate power tools or machinery. It is much safer to leave machining to the next day.
- Get previously owned machines checked by an electrician before using. Replace switchgear with an NVR version if appropriate, and ensure guards are still effective.
- Always keep the workshop tidy, especially the floor. Don't let off-cuts accumulate near machines; throw them in waste bins.
- Useful smaller pieces of wood can be stored by size or species.
- Gardening gloves are useful when lifting heavy, rough sawn boards. Splinters can be nasty, especially from some hardwoods.

Related info

Dust control (see page 20)
Safer woodwork (see page 22)
First aid (see page 23)

Machine safety

- Keep a pushstick by the table saw or bandsaw and use it. Make several pushsticks at the same time, so there is always a spare to hand. Make sure your hands are well away from the blade on any saw and not directly in line when cutting.
- Never operate a saw with the riving knife removed. It is there to prevent timber (lumber) jamming and being thrown backwards. Don't remove the crown guard from a table saw either, unless using a specific jig that completely covers the exposed blade.
- Adjust the guard on a bandsaw each time you cut a different timber (lumber) thickness. It should be only just above timber (lumber) height.
- Always wait until a spinning blade or cutter has reached standstill before clearing off-cuts from the table of a machine, or use a pushstick.
- Never reach inside the bed of a thicknesser (thickness planer) while it is running. If wood should jam, switch off, lower the bed and remove the item with a pushstick.
- Never remove the bridge guard on a surface planer (jointer). Adjust to within 10mm (⅜in) of timber (lumber) moving underneath; with narrow pieces lower completely and slide towards the fence. Keep exposed knives to a minimum.

bandsaws

Unlike most woodworking machines, the bandsaw is designed to cut curves as well as make straight cuts. Set up correctly and fitted with a sharp blade, it excels at cutting deep timber (lumber). You can use a bandsaw to convert large planks or cut delicate dovetails – it is because of its versatility that many woodworkers prefer a bandsaw to a table saw as their first machine. Probably the quietest, it is arguably also the safest machine in the workshop.

The bandsaw

The bandsaw is shown here with door closed and opposite with the door open. The continuous narrow blade runs around the two large wheels and passes through adjustable upper and lower guides to maintain a straight, vertical cut. The blade is tensioned by adjusting a handle or knob that raises the upper bandwheel; this allows for minor discrepancies in blade length. Here, the lower bandwheel is driven via the motor's drive belt, with pulleys to change the operating speed. The frame of the machine is built from heavy gauge steel; it must be stiff enough to withstand any flexing due to the considerable tension on the blade – if the frame flexes it will be impossible to saw accurately.

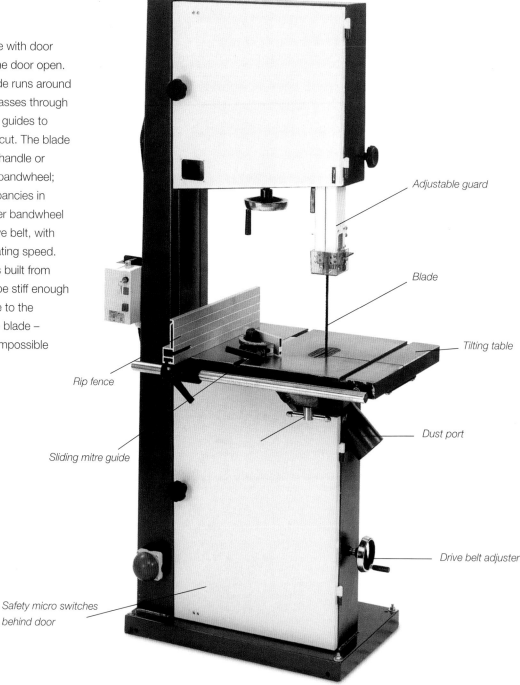

Adjustable guard

Blade

Tilting table

Rip fence

Dust port

Sliding mitre guide

Drive belt adjuster

Safety micro switches
behind door

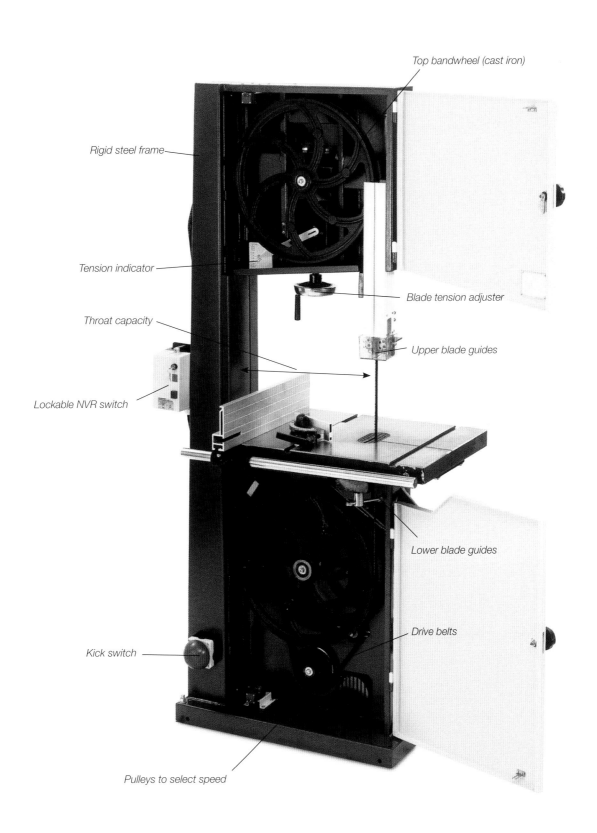

Top bandwheel (cast iron)

Rigid steel frame

Tension indicator

Blade tension adjuster

Throat capacity

Upper blade guides

Lockable NVR switch

Lower blade guides

Kick switch

Drive belts

Pulleys to select speed

woodworking machines

Bandwheels

Upper and lower bandwheels may be cast alloy or steel with rubber tyres for grip and are cambered to enable the blade to track centrally. Ideally the blade should run midway on both wheels – a tracking mechanism behind the top bandwheel can be altered to achieve this. There is usually a brush that rubs against the lower bandwheel to remove sawdust from the tyre surface. Smaller bandsaws sometimes have three wheels and are best avoided – although throat capacity (and so width of cut) is bigger, blades are more likely to break.

Blade guides

Guides are positioned above and below the table and are adjustable laterally and front to back. Depending on timber (lumber) height, the top guide should be raised so it is just clear of the material. A built-in guard provides some protection from the moving blade once upper guides are set. Guide blocks are set within a whisker of the blade on both sides to maintain a true, vertical path. As you feed timber (lumber) into the saw, a rear thrust wheel stops the blade from deflecting.

Related info

Safety (see page 92)
Saws (see page 40)

Tilting table

A bandsaw table is usually cast iron – although on small machines it may be alloy – and should be completely flat for accurate work. It is bolted to a trunnion to enable the table to be tilted up to 45 degrees on some machines and secured with a lever for bevel cutting. A mitre guide slides in a groove machined in the table to enable you to make accurate crosscuts. This guide usually incorporates a protractor scale that can be set for angled cuts, while for straight cuts a fence is locked across the table.

Switches and power

A bandsaw is fitted with an NVR switch; heavier machines also have isolator devices. The motor is located behind the lower bandwheel and usually provides power via a pulley system and drive belts. Many bandsaws have a choice of two speeds, while some are variable.

Adjustable guides above and below the table keep the blade running completely vertical. A thrust wheel (or rod as shown left) supports the blade during cutting.

Bandsaw blades

Blades are welded to form a continuous band and tend to be disposable on all but industrial bandsaws. To increase cutting life, bimetal steel blades have heat-treated teeth with hard tips – the remainder of the blade is untreated so that it remains flexible. Once blunt, these teeth cannot be reset or resharpened. More expensive and less common now, blades with normal (untreated) steel teeth can be resharpened. Certain blades enable you to cut plastics and nonferrous metals.

For optimum performance select a suitable blade (either normal or skip tooth) for the task and tension it properly. For clean sawing, a regular or triangular tooth pattern is popular, with teeth 10tpi or finer. Skip tooth blades have wide, shallow gullets for clearing waste material and are better for sawing dry hardwood.

A reasonable width of blade, for both straight and curved sawing, is 13mm (½in) – this will enable you to cut a radius down to 63mm (2½in). For really tight curves it is possible to fit a really narrow blade of 3mm (⅛in) on some machines. For resawing heavy timber (lumber) the wider the blade the better – widths of 38mm (1½in) are typical on larger bandsaws.

Crucial capacity

When choosing a bandsaw, consider the maximum size timber (lumber) you are likely to cut Depth capacity is the distance from the table to the raised upper blade guides; on large bandsaws this is commonly 254mm (10in), though it may be up to 406mm (16in) on big machines. A bench-top bandsaw may cut 152mm (6in), which could be enough for many woodworkers. Throat capacity is measured from the vertical frame to the blade, with around 380mm (15in) sufficient for most craftsmen.

Fitting a blade

Open the doors and remove guards if necessary. Slacken off the upper bandwheel and carefully remove the old blade. Lift the new blade over the wheels and apply tension with the adjuster. Rotate the upper wheel by hand so that the blade runs freely without touching guides or thrust wheel. Move the adjuster handle so it runs midway on both wheels, a process called tracking. When tracking is completed, set the upper and lower blade guides to support the blade. Check the clearance with a piece of paper, adjust the thrust wheel and finally rotate the blade by hand once again.

Blade care

It takes greater effort to feed timber (lumber) through a bandsaw if teeth are blunt – it is harder to follow the line closely and the blade could snap.

table saws

Literally at the centre of many a workshop, the table saw is a real workhorse for accurate crosscutting, mitre sawing and ripping timber (lumber) and sheet materials. From compact portable bench-top saws for small-scale work to large machines capable of converting full-size panels with precision, the range can be bewildering. Bevel cuts are easily made by tilting the blade, while sliding tables and fences make mitre and 90-degree crosscuts simple and safe. If the machine is correctly set up and has a sharp blade, the resulting cut often needs very little cleaning up.

Related info

Safety (see page 92)
Portable saws (see page 68)
Cordless power tools
(see page 62)
Saws (see page 40)
Layout and workflow
(see page 12)
Dust control (see page 20)
Universal machines
(see page 118)

The table saw

Most static workshop table saws consist of a steel cabinet base enclosing the motor, which drives a rotating circular blade that protrudes through the table and can be adjusted in height. Timber (lumber) is passed across the table, against a rip fence parallel to the blade or held against a sliding fence when crosscutting. On most saws the rip fence is on the right of the blade, with an auxiliary sliding table for precision cuts on the left. You can adjust depth of cut by rotating a handle, which raises or lowers the blade. The blade is bolted to an arbor and can be tilted up to 45 degrees for cutting bevels. Motor rating is typically from 1000 to 2000W on smaller workshop machines, while larger capacity table saws can be more than 3000W. These machines will have heavy induction motors built for extended running time, while cheaper saws tend to have noisy, less powerful brush motors. An NVR switch is fitted at the front, with a highly visible stop button.

Tables are usually cast iron, though budget saws may be cast alloy or steel – cast iron is preferable because it is heavier, has greater stability and is less prone to vibration. The blade rotates through a plastic or hardwood insert, which prevents small pieces of wood jamming during a cut; when changing a blade you remove this insert first. Smaller saw tables generally have a groove for a sliding mitre fence; this may have a locking protractor scale to enable you to make precise mitre and crosscuts easily.

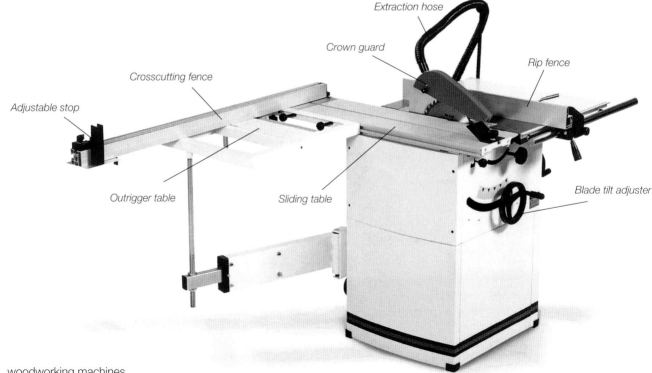

Extraction hose

Crown guard

Rip fence

Crosscutting fence

Adjustable stop

Outrigger table

Sliding table

Blade tilt adjuster

Scoring blades

For an ultra-clean cut on melamine or veneered boards, some panel saws have a smaller scoring blade ahead of the main cutting blade, as shown right. This is generally about 100mm (4in) in diameter, and the teeth protrude just above the table to score the underside of the panel before it reaches the main blade, to prevent any chipping that could otherwise occur. The primary blade diameter may be reduced on some machines if a scoring blade is fitted.

Blades and teeth

Blade rise and fall is controlled by rotating a handwheel – raise the blade so its teeth are just above the wood you wish to cut. A second wheel tilts, or cants, the blade to 45 degrees. A typical table saw for the home workshop will be fitted with a 254mm (10in) diameter TCT blade, producing 75mm (3in) depth of cut at 90 degrees. Note that timber (lumber) capacity is always less when the blade is tilted to 45 degrees. On industrial saws blades can be as big as 406mm (16in).

It is important to fit the correct blade to suit the material you are cutting – formats include positive hook teeth and trapezoidal, triple-chip patterns for cutting sheet materials. Tooth size affects the cut, too, with more teeth creating a finer cut. A coarse rip blade may only have 24 teeth, while one for melamine panels may have 80 teeth or more for very fine cuts. Unless a really fine finish is needed, most woodworkers choose a combination blade suitable for both ripping and crosscutting, with around 40 to 50 teeth being fine for most applications. You should always send a TCT blade to a specialist saw doctor for resharpening.

As a piece of timber (lumber) travels through the saw there is a tendency for it to close up and bind on the blade, which could create kickback. To prevent this, a riving knife is fitted behind the blade and attached to this knife is a crown guard that encloses the exposed teeth – both of these are raised or lowered as you adjust the blade height. Never use a table saw with the crown guard or riving knife removed.

Crosscutting and mitres

Every table saw has a fence of some kind for crosscutting and mitring timber (lumber). In its simplest form this will be a mitre fence that slides in a groove in the table. For accuracy, this should be a reasonably tight fit. More sophisticated is a sliding table to the left of the main machine table – a crosscutting fence fitted to this, with an adjustable length stop and measurement scale, gives greater control and precision. Hold timber (lumber) against the hinged stop and push the fence forwards to make a precise cut; a hold-down device is usually provided so material does not slide about during sawing. For accuracy and convenience the fence will normally have index pins at 90 degrees plus the common mitre angles, so you can fit it to the table speedily without having to check angles. For rip sawing lock the table to stop it sliding forwards. For cutting large panels and crosscutting long boards, some larger table saws will accept a sliding carriage. Fitted to the left of the machine, this offers more support than a sliding table but increases the machine's footprint.

Rip sawing

The rip fence normally extends right across the table for rip sawing and is usually made of extruded aluminium, although it may be cast iron on industrial machines. It slides along a rail at the front of the machine and for accuracy it should lock rigidly. After reading the measurement scale, you set the width of cut by sliding the fence across and locking it off. Most saws include a fine adjuster for nudging the width setting for a precise cut.

Sheet materials need fence support across the full table when ripping, while timber (lumber) only needs support just past the blade to prevent it pinching. For this reason there should be a facing to the rip fence, adjusted to suit the material being sawn.

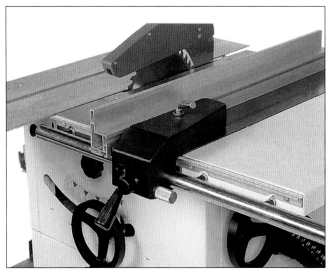

Warning

Always keep a pushstick next to the table saw for guiding narrow timber (lumber) past the blade. When clearing waste material away from the blade, always use the pushstick and NEVER use your bare hands. For narrow off-cuts, switch off the saw and wait until the blade has stopped rotating before clearing the table.

Site saws

The site saw is a no-frills machine, the most basic in this family. Popular on construction sites, it is also known as a contractor's saw since its prime use is ripping timber (lumber) to width. The table is galvanized steel to withstand corrosion and the steel legs may have wheels for mobility. Blade diameter is typically 315mm (12⅜in).

Compact saws

For occasional cutting or where space is limited, a compact bench-top saw may be all you need. Usually constructed with a dense, moulded plastic body and a motor up to 2000W, these saws are sufficiently light to be lifted off the bench when not required and can be transported by car. To save weight, the table tends to be cast aluminium. You can cut material up to 75mm (3in) at 90 degrees with a 254mm (10in) blade. Although mitre and crosscuts are feasible, bench-top saws are best for ripping timber (lumber).

mitre saws

Originally designed for the carpenter or shopfitter working on site, the portable mitre saw is now just as popular in the small workshop – it takes up relatively little space and has largely replaced the radial arm saw. Compact machines are handy for sawing mouldings and smaller timber (lumber) sections, which they can handle with great accuracy. For bigger timber (lumber) and panel sizes the sliding mitre saw is champion – like a radial arm saw, the head travels horizontally, while the blade can be raised for cutting housings. Bevels and compound mitres are made by tilting the blade over – it is possible to tilt to both left and right on more elaborate saws.

Compact mitre saws

On most mitre saws the base is cast alloy, making them tough but light enough to carry or stowaway when not in use. The saw head houses the motor and blade; when cutting the blade is lowered into timber (lumber) that is held against a rear fence. The motor and pivot actions are activated simultaneously by squeezing a paddle handle or lever with the trigger. The blade is almost totally enclosed by a guard, the lower part retracting to expose the spinning teeth as the head is lowered downwards. At the end of the cut, the head returns to its raised position when the trigger and handle are released. For mitre cutting a turntable incorporated in the base is rotated, swinging the entire saw head to left or right. You select an angle on a crescent-shaped protractor scale on the front apron, locking this off with a handle or lever. Common mitre angles of 15, 22.5, 45, 60 and 90 degrees are indented, so these can be selected quickly and accurately. You can swivel the saw round to about 60 degrees left or right on some models.

Typical blade size is 210 (8¼in) or 250mm (10in), revolving at some 4500rpm and driven by a motor of about 1200W or more. The maximum timber (lumber) width (front to back) is limited on smaller saws, although depth of cut can be more than 75mm (3in) at 90 degrees. This reduces to around 45mm (1¾in) when you tilt the blade over up to 45 degrees for bevel sawing – to do this you release a locking knob below the saw head,

selecting the angle on a protractor scale. To accommodate the tilted head, the two-part fence is usually moved sideways.

Precise compound mitres – when bevel and mitre angles merge together, often found on crown mouldings and cornices in cabinetmaking or shopfitting – are difficult to cut by hand, but simple on a mitre saw. For cutting picture frame profiles, where precise mitres are also essential, the mitre saw excels. For safety and precise sawing cramp (clamp) timber (lumber) to the fence or table, which often has a built-in hold-down device. Many mitre saws also have extending steel arms that slide out from the base to act as timber (lumber) supports. As well as a dust outlet, a fabric dust bag is standard on most machines.

> **Related info**
>
> Safety (see page 92)
> Radial arm saws (see page 104)
> Layout and workflow (see page 12)

Sliding compound mitre saws

On a sliding compound mitre saw the head travels horizontally, unlike the vertical movement of a compact saw – steel rails enable you to pull the blade towards you before plunging it down into the timber (lumber). This feature greatly increases timber (lumber) width capacity, which is typically 340mm (13⅜in), and you can also lock the saw head at any point on the rails. A useful feature specific to sliding mitre saws is the trenching function – raise the blade height and you can make precision cuts across the grain. Sliding mitre saws tend to be quite big, with some professional machines equipped with 305mm (12in) blades running at about 4000rpm. At 90 degrees, maximum timber (lumber) depth is around 110mm (4⅜in), or 70mm (2¾in) set at 45 degrees. Much more powerful than compact saws, the motor can be 2000W or more. Increasingly, these machines are equipped with soft-start and electronic braking.

Lasers and lights

Laser cutting guides are increasingly common on mitre saws. You can align a beam with a pencil mark on the workpiece, and the beam can be calibrated to show precisely where the kerf will appear. Handy worklights are standard on some machines, too. A mitre saw is designed to be used at bench height, so it should really be mounted on a floor stand or have its own base. A universal stand with rollers at each end is a useful accessory if you are likely to use the machine for site work, and this will support long timber (lumber) lengths.

Flipover saws

For ultimate versatility and portability, the unique flipover saw is essentially two machines in one. With its built-in floor stand, you can swap between regular mitre saw and compact table saw simply by lifting and flipping over the saw head. You can rotate the cast alloy table for cutting mitres, while the head locks down for table saw use, the blade poking through, A blade size of 254mm (10in) will cut 60mm (2⅜in) timber (lumber) at 90 degrees. Typically 2000W, the same motor powers both functions with switches easy to reach.

radial arm saws

Built for precision crosscutting, the radial arm saw is common in larger workshops although in home shops it is less popular than the smaller sliding mitre saw, which needs less space. Sliding along a fixed overhead arm, the blade assembly is pulled across the table to crosscut; with the saw head swivelled it will cut mitres, while tilting it produces bevels and compound mitres. The radial arm saw is equally suited for cutting housings and tenons with the blade raised above the timber (lumber). Unlike a mitre saw, you can attach a router or sanding drum to certain machines for overhead work.

Radial arm saw essentials

Position a radial arm saw close to a wall, with space on each side for timber (lumber). Although usually mounted on a floor stand, it makes sense to build long workpiece support tables, so you can slide timber (lumber) along as it is cut. Unlike other machines, the saw should be fitted with a sacrificial table – you actually lower the blade to cut slightly into the surface, so there is no tear-out on the material being sawn. This board will deteriorate with repeated mitre or bevel cuts, so make sure it is easy to replace. The rear fence is also sacrificial because the blade passes through it.

Overall arm length dictates timber (lumber) capacity, which is typically 350mm (13¾in) with a 305mm (12In) blade; at 90 degrees depth of cut will be about 90mm (3½in). The power switch is easy to reach close to the handle, while motor rating is about 2000W. The arm is supported on a steel column and can be raised or lowered with a rack and pinion movement. Turning a rear handle adjusts the blade height for making trenching cuts, while for crosscutting hold timber (lumber) against the fence, switch on and pull the handle towards you so the blade travels across the table. The blade is totally enclosed by a guard, which slides upwards as you pull the handle exposing sufficient depth to make the cut. On completing a cut, the saw head retracts automatically along the arm on a spring mechanism. A vertical guard in front of the blade is adjusted independently. For mitre sawing you swing the arm round to left or right, reading the angle from a scale on the column and locking with a lever. Indents at 45 and 90 degrees make common angles fast and accurate. For bevel cuts the blade is tilted and locked.

Related info

Safety (see page 92)
Mitre saws (see page 102)
Table saws (see page 98)
Layout and workflow (see page 12)

scrollsaws

The scrollsaw is the perfect machine for model, doll's house and toy makers. Compact, quiet and safer than other machinery, it is designed for cutting tight curves in sheet materials and thinner timber (lumber). The delicate, narrow blade creates a fine, clean kerf with very little tear-out. Also called a fretsaw, the scrollsaw produces little vibration or sawdust and also cuts metal and plastic.

Saw construction

With its small footprint, you can fit a scrollsaw to a bench or floor stand, although it should be bolted down through the base. The motor is mounted below the table and rarely exceeds 120W. Variable speed is an option on some machines, selected via a rotary dial to suit material density. For sawing timber (lumber) a higher speed of about 1500spm is best, while for soft metals reduce this to around 400spm. The power switch is under the table or on the upper arm.

The vertical blade is held taut between two horizontal arms; a slot in the table allows the blade to be removed once the tension is released. The construction of the arms defines the pattern of the scrollsaw: on a parallel arm scrollsaw these pivot separately, so the blade reciprocates vertically to create a completely perpendicular cut; the C-frame scrollsaw has arms pivoting at the rear and produces a slightly less accurate cut, although the blade does stay completely vertical because the arms create a rocking motion. Similar to a jigsaw, stroke length is about 20mm ($^{25}/_{32}$in).

Timber (lumber) cutting depth on small scrollsaws is limited to around 20mm ($^{25}/_{32}$in), although this can increase to 65mm (2$^9/_{16}$in) on professional machines. Larger saws have a throat capacity up to 610mm (24in). A hold-down device is fitted to the upper arm, which is lowered onto thin material to prevent it lifting off the table during cutting. For bevel sawing tilt the cast iron or alloy table, using the protractor scale to select the angle.

Some machines have a flexible lamp to illuminate the cutting line on a workpiece. Although dust is not much of a problem, a scrollsaw has a flexible plastic pipe that can be moved close to the blade with air supplied by a bellows unit on the base.

Alternatively connect an extractor hose to the dust port.

A quick-release lever simplifies blade replacement When the blade is fitted it is tensioned with an adjuster knob behind the machine – once set there should be no need to readjust this when you change the blade. Blades are a standard 125mm (5in) and come in various patterns, from a coarse 10tpi to a really fine 30tpi. Some scrollsaws are designed to take pin- or plain-end blades, although some will accept both.

Using a scrollsaw

Sawing really tight curves can be limited by blade width; with a spiral tooth pattern there is no need to turn the wood as you cut.

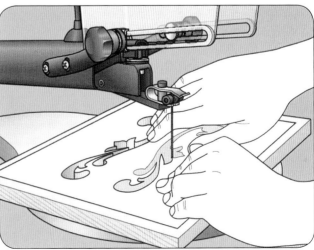

> **Related info**
>
> Safety (see page 92)
> Bandsaws (see page 94)
> Portable saws (see page 68)
> Multi-tools (see page 88)

pillar drills (drill presses)

There will be times when most woodworkers need to drill holes with complete accuracy, whether in timber (lumber), metal or other materials. The stationary pillar drill, or drill press, is a rugged machine for drilling perpendicular to a surface, or at an angle when fitted with a tilting table. It may be bench-mounted or floorstanding, which frees up a work surface and increases depth capacity. Sanding attachments can turn this machine into an extra shaping centre. For occasional drilling work a portable electric drill fitted in a stand is a cheaper option.

Drill basics

A cast iron headstock mounted on a steel column supports the motor, drive belts and spindle assembly. Bolted vertically at the rear, the motor drives the front spindle via a system of belts and pulleys, enabling you to adjust the speed. A hinged cover above the belts gives access if manual speed change is required; before adjusting slacken the belt with a cam lever. A three-jaw chuck is fitted into the spindle, with a Morse taper stem. Every chuck is fitted with a clear plastic safety guard, which can be raised for access. You tighten a drill bit with a key inserted in holes on the chuck, although some machines may be keyless, much like a power drill. The maximum drill bit that can be accommodated does not usually exceed 16mm (⅝in). The drill head is lowered by means of a lever arm or handwheel – its travel can be up to 120mm (4¾in), while throat size is the spindle to column distance. After drilling a hole a spring mechanism raises the drill head again.

The cast iron base has enough weight to make the machine stable, though it is a good idea to bolt it down to a surface. The table may be square or circular, with slots for attaching a machine vice or sacrificial board. Most tables can be tilted for angled work, while rack-and-pinion action offers precise adjustment when raising or lowering. An adjustable stop enables you to set a maximum depth precisely, but first make sure the bit will not foul the table when fully plunged. Swinging the table to the rear of the column increases maximum depth capacity, because the slotted machine base can also be used as a fixed table.

A built-in worklight is provided on some machines. To increase the scope of a pillar drill (drill press) certain models are equipped with a mortising attachment, although maximum chisel size is likely to be less than a true mortising machine offers.

Related info

Safety (see page 92)
Cordless power tools (see page 62)
Electric drills (see page 66)
Drills and bits (see page 56)
Mortisers (see page 110)
Layout and workflow (see page 12)

A pillar drill (drill press) is activated by an NVR switch, with a prominent stop button at the front. Pulley covers also usually incorporate a microswitch, so the machine cannot be switched on when these are open. Pillar drills (drill presses) are powered by very quiet induction motors of up to 650W, so speeds of 600–2850rpm are typical. Larger pillar drills (drill presses) can have 12 speeds, although smaller machines are limited to five – which is enough for most woodworkers. Select a high speed if using a small-size bit for timber (lumber), while with larger bits or when working with metal use a slower speed.

Warning

Never use a pillar drill (drill press) with a centre-threaded bit – this will lift a workpiece off the table or plunge the head down too quickly.

Drill bits

All the bits detailed here can be used in portable electric drills as well as in static pillar drills (drill presses).

Flat (spade) bit Inexpensive and efficient for drilling bigger holes, but more prone to wander than other bits if used freehand.

Forstner bit The ultimate bit for precision drilling, particularly with larger holes, although good quality versions are expensive. The bit is guided by its rim and not a spiral or point, so it is ideal for overlapping holes, awkward grain or angled drilling, For end grain holes, the sawtooth pattern is preferable.

Hole saw With bimetal teeth, for holes in thin metals. The threaded arbor is interchangeable with a range of saws.

Twist bit Designed for metal, but suitable for timber (lumber). Use a centre punch to mark the hole centre before drilling, which will stop the bit wandering.

Lip-and-spur (dowel) bit This bit is designed specifically for timber (lumber). Position the pointed tip on a pencil mark and it will not wander when the machine is switched on. Outer spurs cut a clean hole.

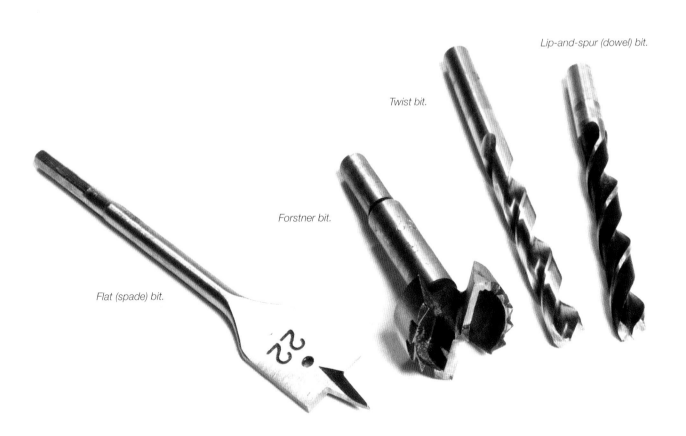

Lip-and-spur (dowel) bit.

Twist bit.

Forstner bit.

Flat (spade) bit.

planers and thicknessers

It is always cheaper to buy rough sawn hardwood than ready-planed boards. By preparing timber (lumber) yourself you can machine it to exact dimensions and have greater control over accuracy and surface finish, but unless you use hand tools for this work you will need dedicated planing machines. Large professional workshops are likely to have a separate surface planer (jointer) and thicknesser (thickness planer), but with limited space it is more efficient to have a combined planer thicknesser (jointer planer), which is available in several sizes.

Planer principles

After timber (lumber) has been sawn into boards, there is a tendency for it to twist slightly as it continues to dry out. The surface planer (jointer) is used to remove imperfections and saw marks and create a clean, straight and flat surface. It should have long, flat machining beds (infeed and outfeed tables) on either side of the cutterblock. As long as the cutter knives are sharp, the result is a smooth timber (lumber) surface that needs minimal tidying up. The face side and edge of a board should always be passed across the surface planer (jointer) first. Remaining surfaces are then planed to the exact width and thickness on the thicknesser (thickness planer). When surfacing with a planer thicknesser (jointer planer), wood is fed above the cutterblock on the beds. When thicknessing it is fed in reverse direction, underneath the cutterblock.

Related info

Planing primary surfaces (see page 166)
Thicknessing wood (see page 173)
Planes (see page 44)
Cordless power tools (see page 62)
Power planers (see page 76)
Layout and workflow (see page 12)
Safety (see page 92)
Dust control (see page 20)
Universal machines (see page 118)

Surface planer

This is used for initial planing of timber (lumber) up to about 204mm (8in) wide. Both tables are cast iron; the infeed table can be raised or lowered with a knob or lever to adjust cutting depth. On heavier machines you can also tweak the height of the rear outfeed table, which should always be in line with the uppermost knives. Timber (lumber) is held against the fence, which can be locked between 90 and 45 degrees. Heavier fences are cast iron, although on lighter machines they are usually extruded aluminium. A bridge guard covers the exposed cutterblock and is adjusted to suit board width and thickness. Note that you cannot thickness timber (lumber) on a surface planer (jointer).

Portable thicknesser

Most compact thicknessers (thickness planers) are portable and designed to be used on a bench top or stand. A powered feed roller passes timber (lumber) under the two-knife cutterblock and out again – it is possible to lock the cutterblock head once set, so there is no movement during thicknessing. Winding an upper handle raises or lowers the bed, adjusting depth of cut, which can be read off an indicator scale. Maximum timber (lumber) width is 330mm (13in), with timber (lumber) thickness no greater than 152mm (6in). Tables are steel rather than cast iron and can be folded up for storage. The machine is normally equipped with a brush motor, rated at some 1800 watts.

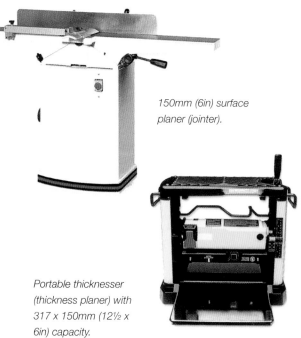

150mm (6in) surface planer (jointer).

Portable thicknesser (thickness planer) with 317 x 150mm (12½ x 6in) capacity.

Planer thicknesser

The planer thicknesser (jointer planer) offers the best of both worlds in one machine, especially if space is at a premium – and with a single motor for both operations, it is generally more economical than investing in a pair of machines. However, a combination machine can be inconvenient if you have to change its operating mode frequently: when swapping from surfacing to thicknessing, first the fence may need to be removed and on some machines you may have to open up the hinged surfacing tables or lift off the outfeed table. The longer the tables the better – a combined length of up to 1520mm (60in) is feasible on larger units. Smaller machines may have cast alloy tables, which avoids rust in a damp workshop, although cast iron tables are heavier and more stable, which is beneficial when planing.

When thicknessing, a lever is used to activate the feed rollers, enabling timber (lumber) to pass under the cutters; for surface planing the feed rollers should not be active. The planer thicknesser (jointer planer) motor needs to be powerful enough to operate both rollers and a cutterblock rotating at around 6000rpm – which means more than 2000W on a compact machine, so ensure your electricity supply can cope. Feed speed is about 3.5m (11½ft) per minute on a budget combination machine, increasing to 8m (26ft) per minute on bigger versions. Maximum planing width on a small machine is generally 254mm (10in), while thickness can be as much as 180mm (7in). Planing depth is adjusted manually using a handle, with an indicator scale providing an accurate guide. The HSS knives can be removed when resharpening is necessary, which should always be done as a pair (or for all three, where fitted). Occasionally a planer thicknesser (jointer planer) will be fitted with disposable carbide knives, which can be replaced easily – these will stay sharp for longer than HSS knives, but you cannot resharpen them.

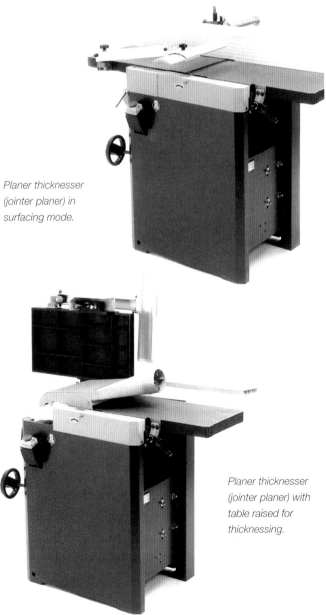

Planer thicknesser (jointer planer) in surfacing mode.

It is more important than on most woodworking machines to have efficient extraction on a planer thicknesser (jointer planer), otherwise chips will tend to clog up the cutterblock and bruise timber (lumber) surfaces. An extractor hood is fitted to every machine, often also incorporating a safety microswitch that acts as a secondary safety device, concealing the cutterblock and isolating power to the motor when the hood is not in place.

Using a push block

Surface planing short pieces or thin timber (lumber) can be tricky because there is a tendency for the material to lift as it passes over the knives, putting your fingers at risk. Keep constant pressure on the timber (lumber) by using a push block made from MDF or an off-cut.

Planer thicknesser (jointer planer) with table raised for thicknessing.

mortisers

A simple machine to use, the mortiser enables you to chop accurate mortises efficiently but needs careful setting up for best results. The mortising head is lowered with a lever and the hollow chisel cuts a square hole in the workpiece. Some machines have a swivel head for angled mortises, while others can be converted to a pillar drill (drill press) by replacing the chisel holder with a chuck and adaptor.

The mortiser

A mortiser may either be fitted to a floor stand or may be small enough to sit on the bench. The mortising head houses a motor and spindle and moves with rise and fall action on a steel pillar, which is bolted to a cast iron base. A chisel and corresponding bit is fitted into the spindle, which is plunged down into the timber (lumber) to cut the mortise. This plunge action is operated manually by means of a large hand lever, with the head lifting again when this is released; on a small bench mortiser the vertical chisel stroke can be 75–102mm (3–4in). When chopping a mortise the workpiece is placed tight against a rear fence, while a hold-down device stops it rising up as you raise the chisel. More advanced mortisers have a sliding table so you can quickly move the timber (lumber) sideways without having to uncramp (unclamp) it, and heavier machines also allow you to adjust the table position from front to back. A depth stop enables you to chop blind or stopped mortises, but fit an MDF or plywood off-cut to the base so that the chisel tip does not contact the metal as you lower it. Motors operate at a single speed and on compact machines are typically rated from 350W.

Related info

Safety (see page 92)
Mortise and tenon joints (see page 200)
Layout and workflow (see page 12)
Drills and bits (see page 56)

Hollow chisels

The mortiser is fitted with a special hollow chisel; as this plunges into the wood, an auger bit rotating inside bores out the waste material. The English pattern auger has two spurs plus spirals for chip clearance, but with no centre tip, while the Japanese auger has just one cutting spur and spiral, but has a centre tip. The Japanese type can be sharpened more easily and cuts faster than the more traditional English bit. Both types of hollow chisel have a slot to help clear the waste material away as they cut – always fit the chisel into the mortiser with its slot at one side, rather than at the front or back. Chisels are sold separately or together with their appropriate bit, in both metric (6–20mm) and imperial (¼–1in) sizes.

Sharpening a hollow chisel

To avoid burning the steel, feed the chisel into the timber (lumber) slowly when mortising. To cut efficiently it is important to keep both chisel and bit sharp. Chisels are sharpened with a fluted reamer, which resembles a countersink bit – it has a projecting pin the same diameter as the auger bit and sits into the end of the chisel. Lock the reamer into a hand drill and turn the handle, removing steel from the inner bevel of the chisel. This creates a burr that is removed by running the chisel on a fine sharpening stone. Several pins are provided with the reamer so they can be changed to suit the chisel size. Use a flat needle file to sharpen auger bit spurs.

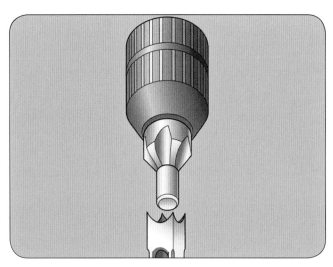

How to mortise

When inserting chisel and auger there must be a gap between them, so the bit is free to rotate – it must not rub against the base of the chisel. A key is used to tighten the auger in the chuck, while a separate hex key locks the chisel and collet securely once set. For a clean through mortise, chop from both faces (just over halfway each time) to prevent tear-out on the bottom.

Mortise cutting

Some woodworkers prefer to make the first cut at each end a few millimetres inside the pencil lines (see pages 200–201) because this is the hardest cut to make and it usually takes several plunges to reach the full depth. The mortise can then be trimmed back to the marked line in later cuts with relative ease and accuracy.

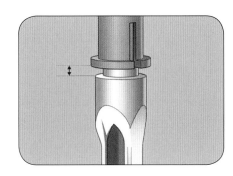

1 Mount the hollow chisel (with the slot at the side) and auger bit in the mortiser. Leave a 2mm (³⁄₃₂in) gap between the lower bit holder and the chisel shoulder.

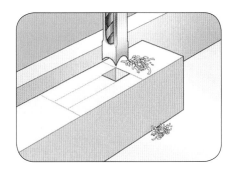

3 Adjust the chisel depth to a little over half the thickness of the wood for a through mortise. With the face side against the fence, cramp (clamp) the workpiece firmly to the table. Lower the chisel smoothly, lifting it often so that chips can escape.

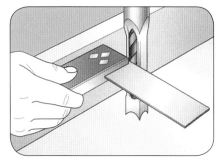

2 Use the side lever to lower the chisel and check that it is square with the rear fence. Untighten the chisel and push it upwards so that it is now tight with the bit holder. Check again that it is still square.

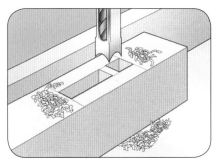

4 Chop both ends first, then move towards the centre with several overlapping cuts. Move the workpiece as necessary, or adjust the sliding table if fitted. Turn the workpiece over, remove waste and repeat the process from the other side, completing the mortise.

sanders

A small stationary sander in the workshop can be surprisingly useful for shaping timber (lumber) components quickly. There are machines designed for both flat and curved shaping, from bench-top disc sanders for cleaning up end grain to bobbin sanders for sanding tight concave curves. The versatile combination sander provides both disc and belt sanding, while for precise thicknessing of wide timber (lumber) with difficult grain, the more specialized drum sander is unbeatable. Always wear a face mask when using a sander and ensure you have adequate dust extraction.

Combination sanders

A good option for a compact workshop, a combination sander incorporates a continuous sanding belt together with an abrasive disc. The two formats mean you can do fast shaping on the belt, finishing off convex curves and end grain on the disc. A belt width of 102mm (4in) or 152mm (6in) is normal on smaller sanders and belts can often be tilted to operate horizontally or vertically. Used horizontally, a fence stops wood from sliding during sanding, while in vertical mode hold the workpiece flat on a lower table. With care, it is possible to grind chisels or plane blades on this abrasive belt. The disc sander has a tilting table for timber (lumber) support, often with a sliding mitre table for bevelled work. Smaller machines have discs of about 125mm (5in) diameter, with larger units up to 305mm (12in). Motor rating is around 250W on smaller units, which are operated with an NVR switch, and dust outlets are provided for connecting an extractor. You can store a bench-top machine out of the way when not required, while larger sanders are stand-mounted.

Related info

Safety (see page 92)
The sanding process (see page 244)
Cordless power tools (see page 62)
Portable sanders (see page 72)
Layout and workflow (see page 12)
Dust extraction (see page 82)
Surface preparation (see page 242)

Disc sander

For small components and forming convex curves, the stationary disc sander is simple and effective. Typically 305mm (12in) in diameter, the flat vertical disc – of steel or alloy – accepts either hook-and-loop or self-adhesive abrasives, which are available from a coarse 40 grit up to 180 grit. Powered directly by a motor of some 500W mounted behind, the disc rotational speed is around 1400rpm. The tilting table allows you to slide timber (lumber) against the rotating disc with a mitre fence, for horizontal or bevelled sanding. Always check the table is square to the disc when resetting.

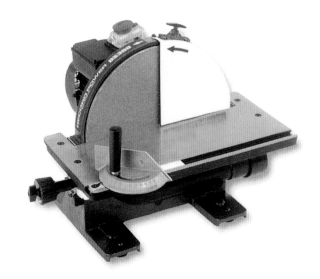

Oscillating sander

An oscillating sander, sometimes called a bobbin sander, is perfect for both concave and convex curves. It consists of a vertical reciprocating spindle fitted with a small diameter sanding sleeve. The motor is mounted in a cabinet beneath a work table – which can be tilted on some machines – through which the spindle extends. Bobbins can be as tiny as 6mm (¼in) in diameter, increasing to 100mm (4in), and may be up to 230mm (9in) long. The stroke length varies from 25 to 38mm (1–1½in). Due to the reciprocating motion there is less clogging of the abrasive, so abrasive sheets last longer. It is best to fit spirally wound, push-fit sanding sleeves – grades from 60–150 grit are available.

Drum sander

For uniformly sanding timber (lumber) to a very specific thickness (such as when working with musical instrument tonewoods), the drum sander offers the greatest precision and is balanced to restrict vibration. A wide abrasive sleeve is fitted around a horizontal metal drum, which revolves above a table, and timber (lumber) thickness is adjusted via a handwheel. A second abrasive belt passing around the table moves the timber (lumber) into the machine, under the sanding drum, and out the other end – feed speed is adjustable to about 3m (10ft) per minute. On the smallest sanders the belt is 254mm (10in) wide, which enables you to sand boards twice this width; because rollers are not enclosed at one end, you can feed work again by flipping the board over and repeating the process. The sanding belt is fitted spirally around the drum, secured with clips. For an excellent finish on wild grain you can fit abrasives down to 240 grit.

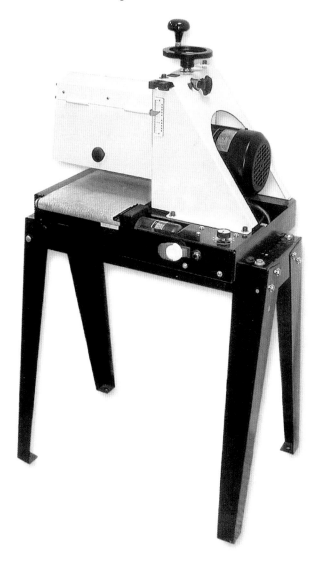

woodturning lathes

Once you have a lathe and a few turning tools, woodturning can be a creative and relaxing way to spend a few hours. A specialized woodworking discipline, turning demands the careful use of tools and lathe. The type of equipment you purchase is governed both by budget and the type of turning you want to attempt. Even with basic skills it is possible to turn small, attractive items from rough timber (lumber) in a short time. A good way to develop these skills is to take part in a short course or join a woodturning club.

Lathe basics

The lathe grips a piece of timber (lumber) securely and rotates it at the same time, so spherical shapes can be created safely with cutting or scraping tools. You can either mount the wood between the two centres (spindle turning), or fit it directly to a chuck or metal disc (faceplate turning). The turning tool is held firmly against an adjustable tool rest, which is mounted on the machine's bed rails.

Related info

Wood characteristics (see page 124)
Seasoning (see page 130)
Bandsaws (see page 94)
Dust control (see page 20)
Safer woodwork (see page 22)
Safety (see page 92)

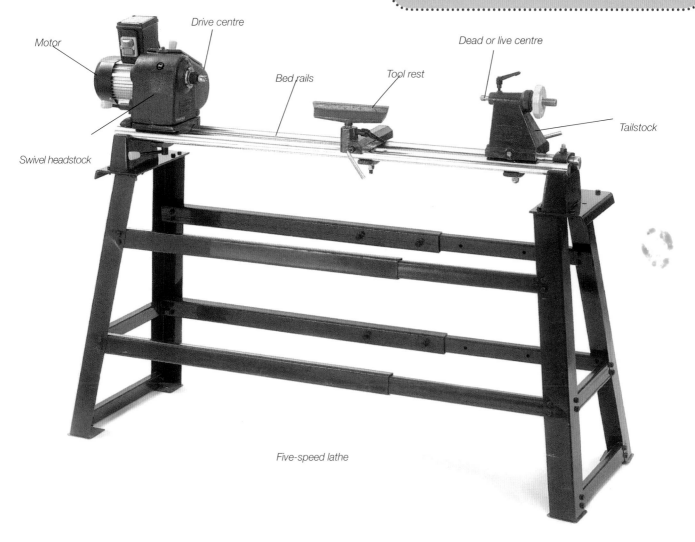

Motor

Drive centre

Dead or live centre

Bed rails

Tool rest

Tailstock

Swivel headstock

Five-speed lathe

A lathe often has several speeds, so you can choose how fast the wood rotates – either manually by adjusting the drive belt between pulleys on a simple machine, or by selecting the speed electronically on more sophisticated models. The larger the diameter of the timber (lumber) being turned, the slower it should revolve. Avoid cheap lathes, since they tend to be poorly made and will make turning a chore – it is better to buy a small, well-built model than a large, inferior machine with greater capacity. Make sure the lathe is at the correct height – your elbow should line up with the lathe centres when standing alongside.

A compact lathe can be mounted on a bench, while bigger machines are fitted to steel or cast iron stands. Build quality and stability is crucial in producing clean, precise turning. The motor is fitted at one end of the machine and drives a spindle that rotates inside a headstock.

Getting started

Green, unseasoned wood is easier to turn than when it is dry and seasoned, so pieces found in the hedgerow – or even logs for the fire – can be perfect for practising tool control. Not only is most of this material free, but also branches tend to be vaguely circular. Check there are no splits or knots, and examine recycled timber (lumber) carefully for nails or screws – if in doubt, don't use it. For spindle turning, pieces roughly 50mm (2in) across are suitable, cut into short lengths. Before powering up the lathe, always make sure you rotate the timber (lumber) first by hand to check if it fouls any mechanical part, before switching on. It is more difficult to achieve a decent finish on seasoned wood, so progress to this after experience with green material. Woodturning suppliers sell pieces of wood ready-cut to shape – called blanks – although you can easily saw your own on a bandsaw.

Spindle and faceplate turning

Most homes contain examples of spindle turning – stair balusters and chair legs are the most familiar. To turn a spindle the blank is supported between two centres: a drive centre in the spindle of the headstock plus a dead or live centre at the tailstock. The drive centre has prongs that you force into the end of the blank with a mallet, while the dead or live centre consists of a pointed cone inserted in a hole at the opposite end.

Drive centre

Lathe specifics

When choosing a lathe consider the following:

Capacity A minimum of 760mm (30in) between centres is required if turning stair balusters or table legs.

Power A 550W (¾hp) motor is ideal for a lathe this size.

Swing For large bowl diameters, choose a capacity of around 305mm (12in).

Speed Four-speed is preferable, varying from 450 to 2000rpm. Variable speed (electronic) is great to use, but can be expensive if a fault develops.

Construction Most important are weight and rigidity, so the machine does not suffer from vibration. If space and budget allow, choose a stand-mounted lathe with solid steel bars or cast iron bed.

To turn a bowl or platter, attach a circular blank (with grain at 90 degrees to the lathe) to the machine with a faceplate or chuck. Screws are inserted through the faceplate into the back of the timber (lumber) – this is a safe method, provided that the wood is sound and its surface is flat. Do not use screws smaller than 5mm (No.10) size, which should be driven tightly into the wood a minimum depth of 13mm (½in). Blanks are normally sawn from timber (lumber) with grain running lengthways so screws are entering side grain and will grip tighter. Always make sure the blank is bigger than the faceplate you are using. For attaching small blanks a screw chuck is convenient, while a combination chuck with four expanding jaws to grip the blank is better for holding bigger pieces of wood securely.

Sharpening turning tools

Turning tools are sharpened on a vertical bench grinder fitted with two abrasive wheels: coarse, general purpose (60 grit) and fine (100 grit). These are usually aluminium oxide, and a good diameter is 203mm (8in), with the finer wheel about 40mm (1⁹⁄₁₆in) wide. The tool rest in front of the grinding wheel should be adjustable, yet rigid when locked. Always wear eye protection when using a grinder of any sort. Unlike general woodworking chisels or gouges, turning tool edges are not honed on a stone after grinding. Depending on the density of the timber (lumber) being turned, tools will often need to be resharpened several times during a work session. Always keep tools sharp and their bevels accurate.

Turning tools

Never use normal woodworking chisels on a lathe because they could break in contact with timber (lumber) revolving at high speed. Woodturning tools are made to cope with these stresses, whether being used for spindle or faceplate work. Blades are usually high speed steel (HSS), with ash or beech handles. Miniature tools are available for use with a small lathe.

A basic full-size kit would typically consist of these tools:
10mm (⅜in) spindle gouge.
Round nose scraper.
19mm (¾in) oval skew chisel.
10mm (⅜in) beading and parting tool.
19mm (¾in) roughing gouge.

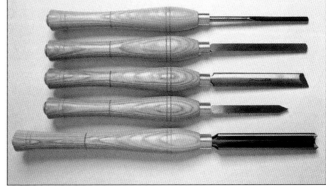

Lathe safety

- Remove loose clothing and tie hair back before operating a lathe.
- Never use a lathe without wearing eye and face protection, preferably a powered respirator, and always use dust extraction.
- Never use a turning tool near the end of the tool rest – stop the machine and move it sideways.
- Don't be tempted to measure or check the workpiece without stopping the lathe.
- Check a wood blank will not foul the tool rest by first rotating by hand.
- Before switching on the lathe, check the tool rest is locked firmly in position on the bed.
- When you switch on the motor stand to one side, just in case the blank flies off the machine.

universal machines

The universal, or combination, machine is a practical consideration for the serious woodworker who tends to work alone; combining several individual woodworking operations in one machining centre means only one function can be used at a time so it is much less convenient in a shared workshop. With a table saw as the hub, most universals combine this facility with a planer thicknesser (jointer planer), sometimes adding a router table or spindle moulder, plus a mortiser. With sufficient space in the centre of the workshop, a universal will probably occupy less total area than the equivalent standalone machines together, and another benefit is that machining tables are at the same height (apart from thicknessing), giving some extra support for large panels and timber (lumber). On the downside, it can be frustrating swapping from one function to another and back again on a universal, especially if you have to remove fences or raise planer (jointer) tables.

Construction and power

The universal is built around a heavy steel cabinet, which houses motors and drive belts. Tables are generally cast iron, although they may be cast alloy on cheaper models. It is possible to take some machines apart reasonably easily so you can get them through a narrow doorway. Most universals are equipped with three induction motors, although a budget machine may be limited to one. With one motor, power is fed to the planer (jointer), saw or thicknesser (thickness planer) by electronic switching or manually selecting the drive belts. With three motors each is switched independently, although for safety only one can be activated at a time. On/off and emergency stop buttons are located around the machine, and should be within easy reach of each function. Check your workshop power supply will be able to cope, because some universals require a 16 amp current. Smaller machines may be available in single phase (240V) or three phase (415V) options; professional universals are generally designed for three phase operation.

Each operation is likely to have an outlet for attaching a dust hose, including a small port on the saw's crown guard, with a larger one beneath the table, an outlet on the spindle moulder safety hood, and another over the cutterblock that is shared by both functions. The diameter of the dust outlets is commonly 100mm ($3^{15}/_{16}$in) and you can attach a small dust extractor with a flexible hose, although a fixed installation is preferable.

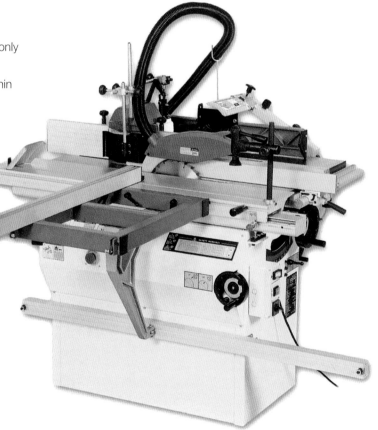

Related info

Safety (see page 92)
Layout and workflow (see page 12)
Dust control (see page 20)

Planing and thicknessing

Timber (lumber) width and depth capacities on universal machines are similar to standalone planer thicknessers (jointer planers). On most universals the surfacing tables are raised to access the thicknessing function. The cutterblock contains two or three knives, with speed varying from 4000 to 6500rpm; thicknessing feed speed is around 8m (26ft) per minute.

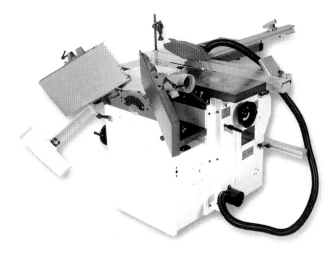

Spindle moulding

Many woodworkers avoid the spindle moulder, but this function is usually included on a universal. Matching cutters are secured in a cutterblock bolted to a vertical spindle, which is raised above the table through an opening. Budget machines may have a single spindle speed, while bigger machines typically have three speeds, from around 3500 to 7500rpm. You can cut tenons easily on the spindle moulder used with the sliding table, or produce profiles in curved and straight timber (lumber) using the fence or shopmade jigs. An extractor hood encloses the cutterblock and you should never run the machine with this removed – and always use the safety guards supplied.

Sawing

The table saw is the heart of any universal, offering crosscutting, ripping, bevel and mitre cutting. A budget machine can have a blade diameter of 200mm (8in), rising to 305mm (12in) on a professional model. Set at 90 degrees, the maximum depth of cut varies from 60 to 102mm (2⅜–4in). Handwheels are used to adjust blade height and tilt for bevel cutting. The full-length rip fence is usually extruded aluminium, and moves along a front rail. When surface planing this fence may need to be swapped around, because it is often common to both saw and planer. There will also be either a sliding carriage or a smaller sliding table, to be used with the crosscut fence provided to enable you to saw timber (lumber) accurately to length. Check the size before buying, because a carriage fitted to the side of the cabinet increases floorspace requirements considerably, although it does give precise control when crosscutting and mitre sawing.

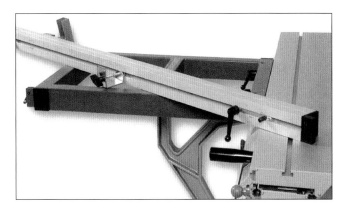

Mortising

A horizontal mortising attachment can be fitted to some universal machines by locking an appropriate mortise cutter into a chuck fitted to the end of the cutterblock. Timber (lumber) is cramped (clamped) to a horizontal sliding table, operated via levers so it is fed into the rotating cutter. Unlike a standalone mortiser such an attachment does not use a square hollow chisel, so the slots cut in timber (lumber) will have rounded ends.

wood and materials

Timber is one of the world's major renewable resources, offering the woodworker a feast of colours, figuring and textures. Each species of hardwood or softwood has its own characteristics and uses, with every piece unique. With such incredible variety, selecting the most suitable material for a project can be slightly daunting. As certain traditional woods become rarer newer species are being established that, over time, will become as familiar and popular. As woodworkers we have a responsibility to look after this amazing resource and use it wisely and creatively.

environmental concerns

We have a responsibility to obtain and use our timber (lumber) wisely, whether it is hardwood or softwood, grown locally or imported from around the world. Whenever possible, using native timber (lumber) does offer several advantages – it is easier to check green credentials, it saves on transport costs and supports local industry. Try using recycled material instead of new, veneers rather than solid wood. We all have ethical questions to consider, but particularly woodworkers.

Vanishing forests

There is no doubt that rainforests are key to stabilizing local and global climate. We rely on the vast tropical forests of Africa, Asia and the Americas to recycle carbon dioxide and create oxygen, essential in reducing the world's greenhouse effect. If we allow these forest zones to become unsustainable, climate change will increase rapidly. We are already seeing the evidence, partly due to deforestation on a massive scale. This is not just due to intensive logging of temperate and tropical forests for timber (lumber), agriculture, oil exploration and mining; increased drought and fires mean even more trees are being lost.

But sustainable use of the world's remaining forest areas is possible. Thankfully, organizations such as the FSC and WWF are heavily involved in forest certification and combating illegal logging, while smaller charities such as Tree Aid work with local communities in Africa to create and manage tree nurseries and sustainable forests. After all, indigenous peoples often rely totally on tropical forests for their income and survival.

Many woodworkers have a simple ethical approach to these complex rainforest issues, refusing to use tropical woods outright and only using native hardwoods. For others, certified timber (lumber) from sustainable sources is an acceptable alternative.

> ### Related info
> World woods (see page 140)
> Sawing techniques (see page 129)
> Buying wood (see page 132)

Sustainable sources

How do we know that the timber (lumber) we are using is actually from a sustainable source and has not been harvested illegally? We have to rely on certification, which guarantees that wood comes from temperate and tropical forests managed responsibly, with saplings replacing trees as they are harvested. The Forest Stewardship Council (FSC) logo gives some assurance that a board comes from a well-managed forest. Ask your timber (lumber) supplier about their stock and their environmental policy, if this is unclear. Encouragingly, an increasing number of independent furniture makers plant trees as part of the deal when commissioned by a customer to build a piece of furniture.

But why use a species that cannot be verified when there may be a suitable alternative wood closer to home? If you choose to work solely with temperate hardwoods and avoid anything vaguely exotic, there is no shortage to choose from. And if you are after a particular colour it is feasible to stain pale, close-grained woods to imitate mahogany or ebony, for instance.

Alternative sources

With both tropical and temperate rainforests shrinking we have to look at alternative sources of timber (lumber). The world is estimated to contain tens of thousands of tree species, but we use just a fraction. Although we may be forced to switch from using a favourite imported wood, either because of increased cost or unavailability, fortunately many less common woods are being introduced so set an example by trying an unfamiliar species. Or why not try a native wood for your next project instead of that more traditional standby from across the world? Both Britain and the United States have a wealth of beautiful, sustainable timber (lumber) that may not be as dramatic as more exotic woods in terms of colour, but can often rival many others when it comes to figure, grain and working properties. And don't forget manufactured boards such as veneered plywood and MDF – for some projects these can be an excellent alternative to certain hardwoods, as well as being more stable and consistent.

Reclaimed timber (lumber)

We can insist on only buying certified timber (lumber) from well-managed, sustainable sources, but is this enough to protect the world's forests? One way to ease the problem – as well as save money – is by using reclaimed wood. Any architectural salvage yard will have piles of floorboards, old doors and so on; it takes time to search through stock to find suitable timber (lumber), but if you can cope with a few nail holes it is quite possible to find excellent, well-seasoned material. Inspect carefully for defects, and before using any machine or power tool check for hidden nails using an electronic detector – a thicknesser (thickness planer) is ideal for cleaning up old boards, but sharp cutters and old nails definitely don't mix! Builders' skips can be a useful source of second-hand timber (lumber), but ask permission before delving in; they may contain discarded material, but this still belongs to someone. And always check for insect attack or rot before taking old boards into your workshop – you do not want to introduce an ongoing infestation problem.

Recycling furniture

Old furniture can be a good source of useful timber (lumber), which will certainly be very well-seasoned and stable. However, even if an item is not a priceless antique it could still be worth something if it is not damaged – if you are in any doubt get a second opinion before you start taking it apart.

wood characteristics

No matter what the species or where it grows in the world, a tree will either be softwood or hardwood. Most hardwoods happen to be dense and heavy, while most softwoods are less dense and lighter, although this is not how specific woods are classified – the differences between hardwoods and softwoods relate to cell structure and formation. Each timber (lumber) species is unique, with its own distinctive characteristics and properties.

Tree of life

The trunk, or bole, is the tree's main stem and supports a crown of branches that bear leaves. A root system anchors the tree in the ground, absorbing water and minerals to sustain it. The trunk carries sap from the roots via the cell system to the leaves. During the life of a tree, a new layer of sapwood forms around the preceding year's growth, so the heartwood expands – the layer of wood formed is known as a growth ring. These layers are much more marked in temperate softwoods and in many species growing where there are very clear annual seasons, so you can often estimate the age of a tree after it is felled simply by counting its growth rings.

Depending on their cell structure, growth rings in hardwoods will be diffuse-porous or ring-porous – grain is more open in ring-porous woods and tighter in diffuse-porous hardwoods, so the latter offers greater consistency when using edge tools. Beech

and maple are diffuse-porous woods with regular cell size and structure, which is due to less marked seasonal changes while the tree is growing. Ring-porous woods, such as hickory or oak, show obvious alternate bands of paler open cells – formed in the fast growth periods of spring and summer – and denser, darker cells – formed in the slow growth autumn and winter periods.

> ### Related info
>
> Seasoning (see page 130)
> Sawing techniques (see page 129)
> Choosing your wood (see page 126)
> Buying wood (see page 132)

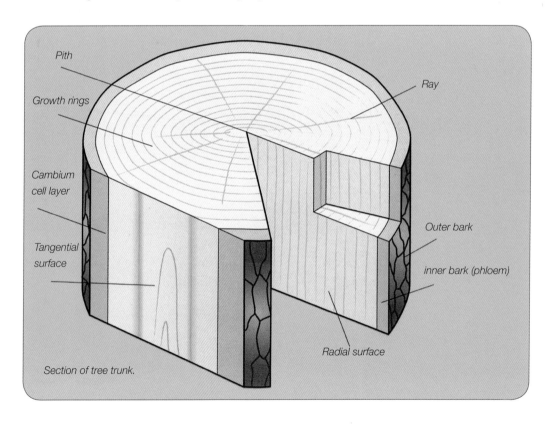

Pith

Growth rings

Ray

Cambium cell layer

Tangential surface

Outer bark

inner bark (phloem)

Radial surface

Section of tree trunk.

Hard or soft timber (lumber)

It is simply inaccurate to describe all softwoods as being light and soft, and all hardwoods as being dense and hard. These expressions are confusing, because woods such as yew and parana pine are both softwoods but are reasonably tough, while balsa is incredibly lightweight but is actually a hardwood. Despite this, most species can be clearly identified and chosen to suit specific woodworking requirements.

Softwoods

Softwoods are coniferous trees (producing cones) with needle-shaped leaves. Their cells are hollow, with food passing through tiny holes in the side walls to reach the leaves. Growth rings are more pronounced than in hardwoods, and sawn timber (lumber) tends to be more pastel in appearance. Faster growing than hardwood, a conifer reaches maturity in roughly a quarter of the time. The softwood family grows mainly in temperate climates, and includes pine, yew, western red cedar, spruce and the giant redwood. The oldest living things on earth are softwoods: bristlecone pines, some of which are estimated to be more than 4000 years old. With a few exceptions, most softwoods are easier to cut and shape than hardwoods, though tools still need to be kept sharp for clean results.

Hardwoods

Hardwoods are broadleaf trees, many of them deciduous (meaning they shed their leaves in autumn) in temperate climates – although confusingly, many hardwood trees in tropical forests are evergreen. Hardwoods tend to be more flexible, the cells longer and more needle-like than those of softwoods. In diffuse-porous woods, such as sycamore, it is hard to detect the growth rings. Familiar hardwoods display a huge variety of colours and grain patterns, and include mahogany, walnut, ash, beech and oak. Hardwoods are naturally more durable than softwoods, and some have natural oils to combat insect attack.

Conifers are the largest and oldest living things on earth.

A magnificent stand of beech trees.

Choosing your wood

When selecting timber (lumber), think carefully about the context in which it will be used. Softwoods grow faster than hardwoods so although they tend to be less durable they are far cheaper. Commonly used in the construction industry, softwoods are treated to increase lifespan and prevent decay. Internal joinery, such as windows and doors, may be hardwood or softwood.

Hardwoods such as oak, maple or walnut will make superb furniture, while ash is still used for traditional sports equipment. Top quality cabinets, boxes and musical instruments often feature exotic woods such as cocobolo or rosewood. Balsa is ideally suited for model making. Iroko, cedar or European oak are all perfect for outdoor woodwork, although make sure you give any exterior timber (lumber) a protective finish and then continue to maintain it. Don't use oak or chestnut if the wood will be in contact with ferrous metals, because these woods contain tannic acid that will react and create black staining. Teak is perhaps the ultimate exterior wood, although it is extremely expensive – cheaper alternatives are iroko and afrormosia.

Certain timber (lumber) species, such as oak and rosewood, include several members within their families. Japanese, English, European, American red and white oaks each have slightly different characteristics. English oak often has beautiful grain, while European oak is straighter and produces less wastage. The mahogany family is huge, if you include substitutes such as sapele and meranti, with timber (lumber) originating from Central and South America, as well as Africa. These woods are often used for door and window frame construction.

Aesthetics

Colour and grain patterns vary extensively across different hardwoods and can make a big difference to the appearance of a piece of furniture. From greens, browns and blacks to vivid reds and purples, it is possible to create stunning effects, especially with veneers. Softwoods are subtler, with pale browns and yellows dominant. Figure can be as dramatic as colour in certain woods. Besides the familiar crown-cut and beautiful quartersawn figuring – English oak's medullary rays, for instance – there are less common effects that are highly sought after. Lacewood (mottled London plane), rippled ash, fiddleback sycamore, and quilted, curly and bird's-eye maple are just a handful of the extraordinary effects available.

Interlocking grain – found in woods such as iroko and sapele – can be difficult to work and creates a striped effect; as you plane one way the grain tends to tear in the opposite direction. Often the only way to successfully prepare such woods

Figure and colour: snake wood, bird's eye maple, English oak, padauk.

is with cabinet scrapers or sanding. Ring-porous woods – such as ash, oak and chestnut – feature open grain (large pores) and are particularly suitable for oil finishes. Diffuse-porous woods – such as box, sycamore and maple – feature finer grain (small pores) and can be polished easily without filling.

Defects and blemishes

Certain defects in timber (lumber) arise from poor sawing – such as fine fractures across the grain caused by the trunk bouncing on the ground during felling – or incorrect seasoning – such as checks or splits – while others – such as knots and burrs – are produced naturally. Many of these defects are not evident until boards are seasoned.

Fungal and insect attack

Avoid storing timber (lumber) in a damp building, because lack of heating and poor ventilation create the ideal environment for wood-boring insects and fungi. Keep a careful check on furniture kept under such conditions, because it can be susceptible to beetle attack. Treat any affected timber (lumber) with an appropriate preservative to eradicate the problem. Exposed constructional timber (lumber) is vulnerable to attack from wet and dry rot, so keep an eye on it. Always try to remove bark from new boards before taking them into your workshop – insects will feed on the sapwood before attacking heartwood.

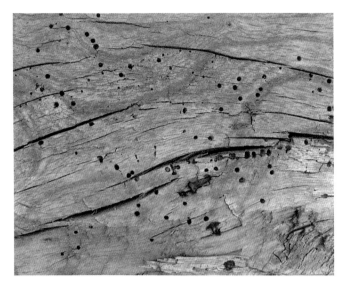

Worm hole damage.

Distortion

Not all distortion occurs after conversion; trees occasionally grow with a natural spiral that results in twisting. Timber (lumber) is also likely to distort if boards are stacked carelessly or sawn badly, with twisting and bowing along the length a common problem. Machining this wood can be tricky because stresses are set up in the timber (lumber). Through-and-through sawing leads to cupping, particularly in outer softwood boards; as this wood dries the faces shrink at varying rates, with a tendency for boards to cup away from growth ring curvature.

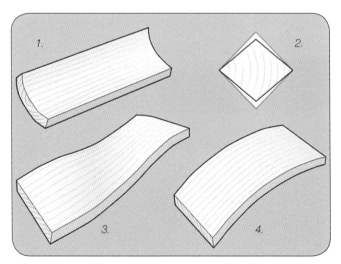

Wood distortion:
1. *Cupping.*
2. *Diamonding.*
3. *Winding.*
4. *Bowing.*

Knots

Knots occur where branches emerge from the trunk of a living tree. They are common in softwoods, with many species graded according to the size and number of knots in a board – small, live knots are usually acceptable. By contrast, knots in hardwoods such as European oak often attract a premium if they are relatively small because they enhance the appearance of the figuring. Dead knots tend to work loose and fall out, so they should be avoided. Depending on the job and timber (lumber) available, it is good practice to plan your cutting to avoid large knots, which will weaken a board. On exterior timber (lumber) the resin weeping from knots can be a problem – always seal knots with a shellac sealer before staining, varnishing or painting.

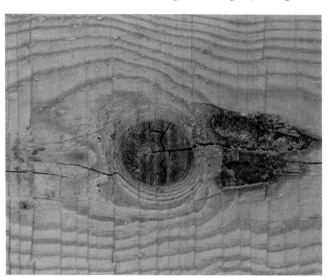

Large knots should be avoided if possible.

Splits, checks and shakes

If seasoning is not managed and timber (lumber) dries too rapidly shrinkage takes place, usually resulting in checks or splits. Splits (also called honeycombing) are found at board ends and edges, while smaller checks can occur over the surface – oak that has been air-dried often suffers from this flaw. Board ends should be sealed with paraffin wax while seasoning to limit end splits.

Similar to checks but larger, shakes are cracks that develop in the tree. If a log is left too long before being sawn into boards, star shakes are likely to appear; these are radial cracks that follow medullary rays and run at 90 degrees to the growth rings. They are caused because the central section of the log is relatively stable, while the outer section shrinks, creating cracks. If the tree is diseased or over mature this can also lead to internal shrinkage, which tends to create heart shakes that radiate outward from the middle. Trees exposed to high winds can often develop ring or cup shakes, where the growth rings part.

A star shake at the end of a log.

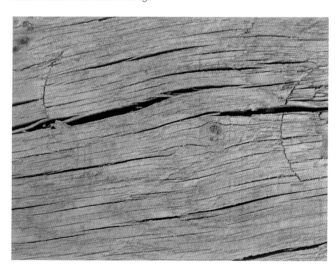

Checks in air-dried oak.

Burrs

These are growths created by the wounds healing after bark has suffered damage and they are sometimes seen on the trunk or branches of a hardwood tree. Burrs often have magnificent, wild grain patterns and are cherished by woodturners especially. Otherwise they tend to be cut into veneers because their random grain makes them a challenge for cabinetmakers.

An oak burr.

sawing and seasoning

After felling, a tree is usually converted into manageable pieces by sawing the trunk into boards. Before the timber (lumber) is ready for use in the workshop it undergoes seasoning, a process that reduces its moisture content to an acceptable level. Without careful seasoning there will be considerable shrinkage problems and the timber (lumber) will be unstable. However, green – or freshly sawn – timber (lumber) is often used in the construction of wood-framed buildings and by woodturners and is allowed to dry naturally, with defects becoming a feature.

Sawing techniques

No matter what the species, quartersawn timber (lumber) is prized by woodworkers for being the most stable. The growth rings run at between 45 and 90 degrees to the surface of a board and often reveal beautiful medullary rays or flecks, especially in oak. At one time timber (lumber) merchants would turn a log numerous times during sawing to obtain the most quartersawn boards possible, but sawmills rarely do this now because quartersawing is a particularly wasteful and labour-intensive technique.

Most sawmills now convert timber (lumber) by through-and-through sawing. By far the most economical conversion method, the trunk is loaded onto a sliding carriage and fed through a vertical band resaw that slices it into uniform boards, starting at one side and gradually working across the log. At the end of each cut the trunk is repositioned and fed through the blade again. Although this technique is less wasteful, only the centre boards are actually quartersawn. Through-and-through sawing can create wide boards, although the very centre of the heartwood is not always usable because it can split easily. One effect of through-and-through sawing is that many species display an attractive flame effect known as crown-cut figure, although these boards are unfortunately much less stable because the growth rings run at a far shallower angle (less than 45 degrees).

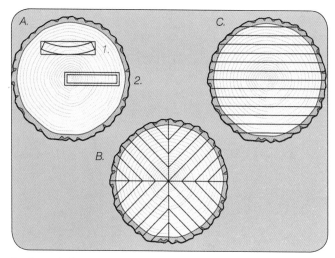

Sawing techniques:
A1. Less stable, but crown-cut figure can be attractive.
A2. More stable, quartersawn from centre of log.
B. Quartersawn boards.
C. Through-and-through boards.

Related info

Choosing your wood (see page 126)
Reclaimed timber (lumber) (see page 123)
Buying wood (see page 132)
Acclimatizing timber (lumber) (see page 162)

Using off-cuts

If you normally build relatively small projects, it is worth getting to know your local joinery company. Such workshops often have usable off-cuts that are uneconomical for them to store, but that can be ideal for the home woodworker. They may be willing to sell you a load for next to nothing.

Chainsaws and bandsaws

Timber (lumber) conversion does not need to involve a sawmill; smaller bandsaw mill operators are increasingly common, particularly in woodland areas. This alternative to large-scale timber (lumber) production enables a tree to be converted where it is felled, instead of first having to be transported to a sawmill. Smaller trunks can be sliced with a chainsaw mounted on a steel framework, the blade moving horizontally along the timber (lumber).

Larger trees may be converted with a bigger capacity bandsaw mill; the trunk is loaded on to a carriage hydraulically, and a diesel-engine driven bandsaw is then fed along its length mechanically.

Seasoning

Every tree contains a huge amount of water when felled, but reducing this must be carefully controlled to avoid defects that make the wood unusable later. Sawn timber (lumber) is described as having a moisture content, which is the weight of moisture given as a percentage of the weight of the wood when totally dry. This means that for a board having 15 per cent moisture content there is 1.5kg (3.3lb) of water to every 10kg (22lb) of dry timber (lumber). The most basic way to monitor percentage is by calculation; weigh a small off-cut of wood, dry the sample in an oven and then weigh it again. An electronic moisture meter is much easier – pin electrodes are inserted into the wood to measure electrical resistance and give accurate readings. Timber (lumber) is wettest when freshly sawn with a moisture content of about 30 per cent, which is called fibre saturation point. No matter whether wood is air- or kiln-dried, its moisture content fluctuates according to temperature and humidity; timber (lumber) gains or loses moisture as it acclimatizes.

Air-drying

Air-drying is a very slow process and means storing the sliced boards methodically over a period of months or even years. The advantage for woodworkers is that this natural drying method is cheap, and boards can be stacked outdoors, although they must be under cover. Stickers, or spacer battens, are placed evenly between boards as they come off the saw and are stacked, to allow air to circulate freely. Water is drawn to the surface by capillary action and evaporates, so moisture content is reduced to around 20 per cent by air-drying. The thicker the board the slower the seasoning process, but for each 25mm (1in) of board thickness you should allow one year.

In temperate climates the driest wood will still contain around 16 per cent moisture content when carefully seasoned, even after several years. For furniture in centrally heated and air-conditioned buildings this needs to be reduced to about 9 per cent for the timber (lumber) to remain stable. You can reduce moisture content gradually in a heated room – although stack correctly to avoid distortion (for further information see page 162). A moisture content of about 14–18 per cent is better for exterior woodwork, such as for framing or sheds. Don't be tempted to use timber (lumber) with a higher level than this, because this will give rise to dry rot.

Kiln-drying

Kiln-drying is much faster than air-drying, and enables you to control moisture content for a specific purpose, but it is a more expensive process. Kiln-drying involves stacking the sawn boards on a trolley and moving them inside a dehumidifier kiln for several days – the process is carefully controlled so that moisture is not extracted too rapidly. Occasionally surface splits can appear on the boards, an indication that kilning was too fast and has left the outer surface too dry, while the inner section was still too wet.

This is called case hardening, and the timber (lumber) will distort when you slice the boards by machine sawing, while in a very warm workshop the damp inner fibres will react, causing the boards to warp.

If the timber (lumber) will eventually be situated close to a source of heat, such as a radiator, moisture content should ideally be down to 9 per cent; for internal joinery or furniture, 11 per cent is normal. When building furniture or joinery items it is always wise to machine your timber (lumber) slightly oversize, and then allow it to acclimatize in a heated workshop. After a few weeks, components can be machined to exact dimensions. Kiln-dried boards should always be stacked indoors; if they are stored outdoors they will absorb moisture from the air and the moisture content will increase.

buying wood

Entering a sawmill or timber (lumber) yard for the first time, it may seem like another world. Some are quite traditional, full of native woods with familiar names, others have stacks of unrecognizable imported boards with strange names. Buying timber (lumber) can be bewildering initially, so persuade a knowledgeable woodworker to accompany you to point out the pitfalls, if possible.

Softwood buying

It is relatively easy to buy softwoods because most imported timber (lumber) is square-edged, so you can see exactly what you are getting. Prepared timber (lumber) makes it easy to spot defects, although you are paying for the machining process. It is sold in stock sizes, with the planing having reduced the original board width and thickness by up to 6mm (¼in). Unless you go to a specialist timber (lumber) supplier, it is likely that rough sawn timber (lumber) from a DIY store or builders' merchant will be a construction grade and inferior to prepared softwood – for best quality timber (lumber) you need to specify a joinery grade.

In Britain softwoods are generally priced by the cubic metre or metre run. Confusingly, hardwood suppliers are far more traditional and this is still sold by the cubic foot or foot run. Width and thickness may seem to have been forgotten when timber (lumber) is priced by cubic or length measurements, but the timber (lumber) supplier will have taken this into account.

Related info

Sawing techniques (see page 129)
Defects and blemishes (see page 126)
Reclaimed timber (lumber) (see page 123)
Cutting lists (see page 157)
Planning and costing (see page 159)
Workshop storage (see page 18)

Hardwood buying

Buying hardwoods is more difficult: boards are less likely to be square-edged, although they will be a uniform thickness, often increasing in ½in or 1in increments for the same species. When possible, select boards yourself at a timber (lumber) yard, so you can examine them for defects. Take along a cutting list and pencil so you can mark roughly what you are likely to need. Unless you are buying a large quantity of timber (lumber) it makes sense to take a vehicle so you can get the pieces home easily; take a saw to cut boards into manageable lengths because the supplier may charge extra to do this for you. Always check your cutting list very carefully, allowing extra length when cutting to be safe. If you cannot transport the wood yourself, most timber (lumber) yards will deliver for a fee.

Many timber (lumber) yards or sawmills will prepare boards to your cutting list if you don't have a planer (jointer) or thicknesser (thickness planer). You may have to wait a day or two and you probably will not be able to make decisions around defects or grain pattern, but this facility could save time and effort if you normally work only with hand tools. A mill may charge by the hour defects when estimating the amount of hardwood needed for a task. This can be very high in woods such as oak and yew – you should allow up to 20 per cent extra on the quantity actually required.

What to look for

It makes sense to check how the sawmill actually stacks its timber (lumber) before buying. Not every yard is particularly tidy and boards loaded carelessly on top of each other without being correctly in stick (stacked with spacers) can lead to bowing lower down in the stack. Sight along the edges of the board and do not be tempted to buy if it is bowed, no matter how attractive the grain – it will not straighten once fibres have distorted during seasoning, since the cell walls will be damaged.

When buying wood it is likely that other people will already have chosen the best boards and the timber (lumber) remaining may well contain defects that could create problems when you start to use it. When sorting through planks always examine both faces – the reverse of a decent-looking board may show bigger knots or wild grain, limiting its use for furniture or perhaps structural work. Most suppliers will make deductions for major flaws, which may be worthwhile if the grain is attractive. Most timber (lumber) with obvious splits and shrinkage defects should have been marked in some way or removed from the stack.

When examining a board, remember that the more parallel the grain, the less it will shrink – but don't rule out boards with wild figure if you find them because these can be stunning if you are confident of your skills in working with them. Sometimes you may have to saw a wide, defective board that has amazing figure into narrower widths and glue them back together; done carefully, this will reduce the stresses in the wood but you will still have a wide board. Often a waney-edge board will taper along its length so to calculate the volume of wood the supplier will measure across both ends and take the average width – lengths and widths are the same whether you buy square or waney-edge boards. If a wood species is new to you but attractive, it may be worth buying a small piece first and experimenting before investing in more boards.

Caring for stored wood

Paint the ends of boards to stop them drying too rapidly and developing splits, a potential problem in heated shops. Where dampness is a problem, install a small dehumidifier to reduce humidity and maintain equilibrium. Cover up timber (lumber) if it is likely to be subjected to direct sunlight; its colour may darken – cherry, for example, will darken in sunlight – or it may fade, depending on the species.

veneers

Veneer is simply a very thin sheet of solid wood evenly sliced off a log. Both hardwoods and softwoods are cut into veneer form, providing an economical and responsible way of using some of the world's most stunning species. Thicker constructional veneers are laminated together to form plywood, while decorative veneers are usually glued to a backing board or groundwork to prevent movement. Veneer gives the woodworker the chance to use exotic figure and grain without the problems and costs often encountered with the solid wood.

Veneer

Available in either familiar woods with regular, straight grain or the most outrageous burrs or gorgeous quilted figuring, veneer can be quite bland or used to make a statement, particularly in fine furniture. With some exotic and rare species veneer is the only form in which the wood can be obtained. Decorative veneer thickness is 0.6mm (¹⁄₆₄in), while constructional veneer is thicker, from 1 to 3mm (¹⁄₃₂–¹⁄₈in). With a well-tuned bandsaw it is even possible to cut your own veneer.

Related info

Manufactured boards (see page 136)
Shaping and bending (see pages 224–239)
Types of adhesive (see page 220)

Forming veneer

There are two methods of forming veneer: slicing and rotary cutting. The most common process is slicing, where the log is mounted onto a huge guillotine. First the timber (lumber) is sawn square and saturated in hot water. The blade cuts a flat, even slice vertically along the length of the log, before retracting; as it does so the timber (lumber) advances for the next cut. As the leaves are sliced in sequence across the log, identical figure is repeated on adjacent leaves, which are stacked as they come off the guillotine. After drying and stacking, the leaves are taped together as a bundle to retain the sequence – these bundles (24 or 32 leaves) make up the complete log. Certain woods, such as oak, may be quartersawn instead of flat sliced to get the best figuring. Rotary cutting involves mounting the log on what looks like a huge lathe, after it has been de-barked and steamed. A blade is fed against the rotating log to produce a continuous sheet of wood veneer, reducing the diameter of the log as it goes. Curl veneers and other patterns are formed by changing the position of the log.

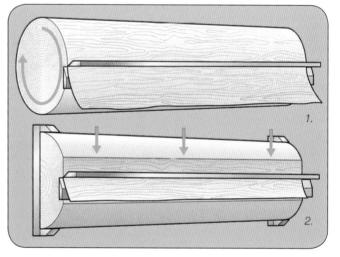

Cutting veneer:
1. Rotary cutting produces a continuous sheet.
2. Flat slicing with a guillotine.

Figure and grain

Veneers are often cut from clean logs that give uniform grain patterns but without dramatic figure. Rotary cutting is an economical way to produce straightforward balancing (backing) veneers, or constructional veneer for plywood construction. Sliced veneers will be more representative of the wood, but the straight grain may be unexciting if quartersawn. Crown-cut veneer is more interesting, creating a wider leaf than quartersawn. At the other end of the scale are exotic and figured veneers. Converting a hardwood or softwood log commercially into veneer form may be expensive, but when sliced some trees reveal distinctive texture and figuring that can be jaw dropping. Certain woods display streaks and shimmering bands that tend to be subtler than medullary rays, although still run at right angles to grain direction. Fiddleback, quilt, curl, pommele and mottle describe some of these beautiful patterns, which are more common in certain woods such as maple and sycamore. Burrs are usually cut into veneers, although the width can be quite limited. Often the figure and real beauty of a species can only be found in veneer form, so it should certainly not be regarded as a cheap method of furniture making.

Buying veneer

In the trade, veneer is sold by the complete bundle (also known as a flitch) although some specialist suppliers will sell an individual leaf. When buying, inspect a bundle to check leaves have not been interfered with – it should be obvious from the sequence if any have been removed or misplaced. Check leaves have been sliced properly by holding one up to the light; daylight will reveal if it is thinner on one side, which indicates poor cutting. Also examine the colour, which should be even across the leaf unless there is sapwood variation. If the veneer has been badly stored the edges may be discoloured. Back in your workshop, make sure you keep the veneers out of direct light, and preferably covered. If possible store them flat in a cool and dry environment. It makes sense to number each leaf of your bundle in the correct sequence or it may be very difficult to match them exactly to get them in the correct order again once they have been spread out.

manufactured boards

Manufactured boards or sheet materials offer several advantages over solid wood. They come in standard sizes and thicknesses, and are stable, economical and widely available. Veneered boards offer the woodworker a greener alternative to using certain hardwoods, with fewer environmental concerns. All manufactured boards, but MDF and chipboard in particular, require very sharp saws for clean cutting; the high resin content will blunt saw teeth and plane blades faster than real wood. It is essential to wear a good dust mask when machining these materials, because breathing in their fine dust is harmful over a period of time.

Related info

Veneers (see page 134)

Workshop storage
(see page 18)

Dust control (see page 20)

Design principles for boards
(see page 152)

Cutting manufactured boards
(see page 177)

Flexible plywood (top) and birch multiply (bottom).

Plywood

Plywood is a versatile sheet material and comes in several grades, from shuttering that is used for concrete formwork and temporary structures to the finest quality birch plywood produced in northern Europe for furniture making. Plywood is made up of several layers of thick veneer laminated together to form a sandwich – there should always be an uneven number for stability. The grain of each veneer runs at 90 degrees to the next layer, and this alternating pattern creates a rigid, stable material. Either softwood or hardwood can be used for the outer surfaces.

Cost fluctuates widely, depending on the woods used and the plywood quality. When it is used for concrete moulds, shuttering plywood is coated with a release agent so boards can be removed easily once the concrete has cured. Marine plywood is used extensively in boat-building because it is water resistant and so perfect for exterior work, and both internal and external grades of birch plywood are common. Regular plywood may be up to 25mm (1in) in thickness, but for model making you can obtain special aircraft plywood, which is about 1mm (1/32in) thick. Plywood is less popular now than MDF, which tends to be more uniform and is cheaper to produce.

MDF

Medium density fibreboard (MDF) is an inexpensive, multipurpose sheet material, used for building everything from joinery to toys and furniture. In the construction trade, softwood is frequently replaced with painted MDF for internal mouldings such as architrave and skirting. It is made from tiny particles of wood compressed together with heavy-duty resin, and is particularly stable and cuts well, although a hardpoint saw is advisable. MDF dust is unpleasant to breath in, so always wear a face mask when sawing or sanding. You can shape and mould edges just like solid wood, though use TCT cutters because MDF will dull HSS tools rapidly. Edges of veneered boards are often lipped with matching hardwood, giving the impression of a real wood surface. MDF makes a good solid base for veneering and with appropriate sealers a fine finish is possible by spraying or painting.

Available in thicknesses of 2–70mm (3/32–2¾in), plain MDF has a rather bland appearance. Veneered or melamine-faced MDF is another story, however, and it is used extensively to make high quality furniture. Popular with contemporary furniture makers, coloured MDF is similar to regular MDF except the colour runs completely through its thickness – before boards are compressed the wood fibres are mixed with organic dyes – so you only need apply a clear finish to protect the surface. For kitchen units and bedroom and office furniture, melamine-faced MDF is tough and hygienic and comes in a variety of plain and wood-effect finishes. For external and construction work you can buy moisture resistant (MR) and fire resistant (FR) grades. Where curves are needed in furniture or joinery work, flexible MDF is a simple solution; about 6mm (¼in) thick, it is rigid once fixed in position.

Veneered boards

Far more stable than solid wood, veneered boards are also more economical and often just as convincing. These days they are mostly MDF veneered on both sides with either hardwood or softwood, which tend to be more common woods such as oak, ash, maple, beech, American black walnut, sapele, cherry, birch and pine. Only suitable for internal use, thicknesses range from 4 to 26mm (⁵⁄₃₂–1¹⁄₃₂in), with board lengths up to 3050mm (10ft). Depending on thickness there is usually a choice of both sides having decorative surfaces, or a cheaper balancing veneer on the reverse if only one side will be visible. Often there will be crown-cut veneer on one side with quartersawn on the other. Edges of veneered boards are particularly vulnerable, and boards need careful handling and sawing to stop tear-out on the back.

wood and materials

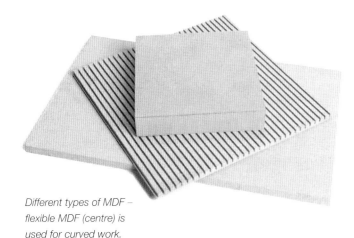

Different types of MDF – flexible MDF (centre) is used for curved work.

Veneered MDF: pine (top), American black walnut (centre), maple (bottom).

Coloured MDF is a consistent colour throughout its thickness.

Chipboard

To make this cheap composite material, wood chips are mixed together with synthetic resin and compressed to form chipboard, or particle board. Common grades have a smooth outer surface of fine particles, with coarser chips sandwiched in the centre. Chipboard is popular in the construction industry where it is used for roofing and flooring, usually tongued and grooved along the edges. This jointing method increases strength and prevents the boards sagging, which is important for sturdy flooring. Moisture resistant grades can be used externally. Chipboard is weaker than plywood and without sufficient support will bow, so avoid using long lengths for shelving. Boards can be reinforced by gluing on solid wood lipping, which will also conceal the unattractive edges. Melamine-faced boards are used for furniture and shelving – on these panels the edges are usually concealed with iron-on plastic lipping. Laminated kitchen worktops usually have a chipboard core and may be up to 50mm (2in) thick.

Hardboard and blockboard

Like MDF, hardboard is a compressed fibre product made from resin mixed with minute wood particles. It is still used as back panels for inexpensive furniture – in which case a white-sprayed surface is common – although MDF is more robust and a better alternative. Uneven floors are often covered in hardboard before laying laminate or carpet. Hardboard thickness is more limited than with other sheet materials, ranging from 2 to 6mm (³⁄₃₂–¼in), Tempered hardboard is a tougher material impregnated with resins and oils, with better moisture resistance. Pegboard is hich is used with special hooks for tool storage. To prevent hardboard buckling after it is nailed to a framework, always condition it first. Lay the sheet down and brush water on the rear face, then leave it to dry before fixing. The sheet will expand and pull up tightly once it is dry.

19mm (¾in) chipboard.

6mm (¼in) hardboard.

Manufactured boards come in standard sheet sizes, irrespective of type and thickness.

Although less popular these days, blockboard is still used in shopfitting and cabinet construction. Narrow lengths of softwood form the inner core, covered with veneers at 90 degrees to the grain of the strips. Far eastern woods are typically used for the veneer – although additional decorative veneers such as oak or ash are used for quality work, with grain running parallel to the grain of the inner core. Blockboard quality varies, with the inner core softwood often not adequately dried – once this is dry, a core strip pattern may show on the finished surface of the panel, which is known as telegraphing. On inferior quality blockboard the inner strips don't always meet together tightly, which may not be obvious until you come across a gap when sawing.

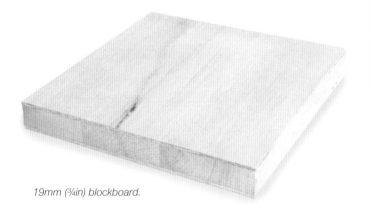

19mm (¾in) blockboard.

Planning ahead

When using manufactured boards you can save time and effort by planning the cutting beforehand, so it often pays to make an accurate drawing for a project first. Some larger DIY stores will cut sheets to size, although they may need advance warning if you have a large cutting list of several different sizes. Specialist suppliers almost always need a few days notice, but the results will be worth the wait. If you are unsure of the finished material sizes needed, you can usually order full-size sheets to be delivered, then saw these to size on a couple of trestles.

Sheet size

The standard size for a manufactured board is 2440 x 1220mm (8 x 4ft), irrespective of the material, with thickness being the main variable: common thicknesses are 6mm (¼in), 12mm (½in), 18mm (¾in) and 25mm (1in). Certain MDF sheets are available in lengths of 3050mm (10ft), with widths of 1200mm (4ft) or 1525mm (5ft).

Melamine-faced MDF is produced up to 2600mm (8½ft) long, with widths of 2050mm (6¾ft). Too large for most home workshops to handle, these giant boards are usually cut first on a vertical wall saw before dimension sawing – most suppliers will cut manufactured boards to match your cutting list.

Using manufactured boards

For clean, accurate edges it is best to cut manufactured boards with a portable circular saw or table saw. Fit a TCT blade, ideally with 60 teeth if sawing veneered panels, and cramp (clamp) a guide rail or batten to the board for straight, accurate cuts with a router or saw. Always use dust extraction when machining manufactured boards and wear an effective face mask. If trimming solid wood edge lippings flush with a veneered board, take care to avoid planing through the face. When gluing heavy lippings to sawn edges, do not rely solely on adhesive for strength – using biscuits or a loose tongue prevents lipping slippage as you cramp (clamp) it, while increasing the gluing area. Plywood, MDF and chipboard can all be jointed with mechanical fasteners, making it straightforward to build carcasses. This hardware tends to rely on holes drilled accurately at 90 degrees, so use a drill stand or pillar drill (drill press) where necessary. A cheaper method is to use plastic connector blocks, where appearance is not important.

world woods

This section covers six common softwoods and 24 hardwoods.

Softwoods

Cedrus libani
Cedar of Lebanon

Origin Middle East, Europe
Sustainability Unlikely to be certified
Used for Box linings and drawers
+ Strong aroma repels insects,
wide boards
– Fairly brittle, expensive

Larix decidua
Larch

Origin Europe
Sustainability Not endangered,
some certified sources
Used for External joinery
+ Durable outdoors, wide boards,
distinctive straight grain
– Prone to splits and knots

Pinus sylvestris
European redwood

Origin Europe, North Asia
Sustainability Plentiful
Used for Joinery, house construction,
furniture
+ Cheap, plentiful, easy to work
– Knots can be a problem

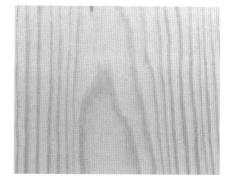

Pseudotsuga menziesii
Douglas fir

Origin Canada, western USA, UK
Sustainability Not endangered,
some certified sources
Used for Joinery, house building
+ Straight grain, fairly strong and water
resistant, sometimes knot-free
– Can be brittle and prone to splintering

Taxus baccata
Yew

Origin Europe
Sustainability Frequent in churchyards
Used for Veneer, musical instruments,
furniture, archery bows
+ Gorgeous colour and grain, bends well
if straight-grained wood
– Expensive due to very high wastage

Thuja plicata
Western red cedar

Origin Europe, North America
Sustainability Not easily regenerated,
supplies of quality wood low, some
certified sources
Used for Musical instruments, roof
shingles
+ Naturally durable and easy to work
– Dust can be irritating

Hardwoods

Acer pseudoplatanus
Sycamore

Origin Europe, Western Asia
Sustainability Not endangered, but
some timber (lumber) certified
Used for Furniture, musical instruments,
woodturning, kitchen utensils
+ Fine grain, few defects, easy to bend
– Surface can burn when machining, not
as hard as maple

Acer saccharum
Bird's-eye maple

Origin Canada, North America
Sustainability Certified timber (lumber)
available, but not under threat
Used for Furniture, musical instruments,
veneered panels
+ Exotic figure prized for cabinetmaking,
more common as veneer
– difficult to work, tough on edge tools,
expensive

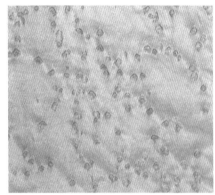

Betula pendula
Birch

Origin Europe, North America
Sustainability Not endangered
Used for Plywood, utility furniture
+ Fine, straight grain, bends well
– Unless figured is rather bland, although
stains well

Buxus sempervirens
Boxwood

Origin Europe
Sustainability Rare, but not certified
Used for Woodturning, tool handles,
musical instruments
+ Fine, smooth, close grain,
attractive colour
– Restricted to small diameter branches,
limited quantities make it expensive

Chlorophora excelsa
Iroko

Origin Africa
Sustainability Low risk, hard to find
certified timber (lumber)
Used for External joinery, garden
furniture, boat-building
+ Strong and fairly stable, oily
and durable
– Interlocking grain, unpleasant to work,
dulls blades quickly

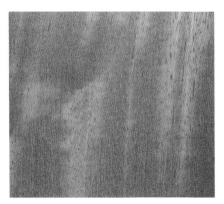

Dalbergia
Rosewood

Origin India, South/Central America,
Mexico
Sustainability Endangered, some Indian
timber (lumber) plantation-grown
Used for Musical instruments,
cabinetmaking, tool handles, veneer
+ Dense, beautiful colour and figure
– Endangered, prone to fine surface splits

Dalbergia retusa
Cocobolo

Origin Central America
Sustainability Vulnerable, certified timber (lumber) rare
Used for Woodturning, inlay, musical instruments, veneer
+ Amazing colour and grain, water resistant
− interlocking spiral grain is common; very expensive

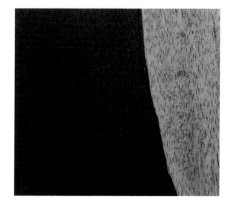

Diospyros celebica
Ebony

Origin Indonesia
Sustainability Vulnerable, certified timber (lumber) rare
Used for Cabinetmaking, woodturning, inlay, musical instruments
+ Extremely hard and dense
− Seasons slowly, risk of splitting, blunts tools rapidly, rare and very expensive

Entandrophragma cylindricum
Sapele

Origin Africa
Sustainability Status varies, but certified tirnber (lumber) scarce
Used for Furniture, flooring, doors, windows, plywood, veneer
+ Wide boards, not too expensive
− Interlocking grain can tear

Fagus sylvatica
Beech

Origin Europe
Sustainability Not endangered, some certified sources
Used for Workbenches, woodwork tools, commercial furniture
+ Good for bending, inexpensive
− Shrinkage a problem

Fraxinus excelsior
Ash

Origin Europe
Sustainability Not endangered
Used for Boat-building, furniture, sports equipment, veneer
+ Strong and flexible, ideal for bending
− Grain can tear when planing, prone to splits

Guibourtia demeusei
Bubinga

Origin West and Central Africa
Sustainability Not endangered
Used for Cabinetmaking, tools, musical instruments, veneer
+ Striking colour, can be highly figured, stable when dry
− Grain can be interlocking, blunts edge tools easily

Juglans nigra
Walnut, American black

Origin Canada, USA
Sustainability Not endangered, certified timber (lumber) easily available
Used for Gunstocks, musical instruments, furniture, joinery, veneer
+ Fairly straight-grained, easy to work
– Dents easily

Juglans regia
Walnut, European

Origin Europe, parts of Asia
Sustainability Not endangered, though timber (lumber) sparse and not certified
Used for Furniture, box making, woodturning, veneer
+ Beautiful grain, figure and colour, easy to work
– Very expensive, risk of insect attack

Ochroma pyramidale
Balsa

Origin West Indies, Central America
Sustainability Not endangered
Used for Model making, carving
+ Excellent to carve with sharp tools, good for buoyancy aids
– Very soft and weak, will crush easily, expensive

Platanus acerifolia
Lacewood (plane)

Origin Europe
Sustainability Not endangered
Used for Furniture, veneer
+ Stunning figure when quartersawn
– Roupala and silky oak may confusingly be sold as lacewood

Prunus serotina
Cherry, American

Origin North America
Sustainability Not endangered, plenty of certified timber (lumber)
Used for Furniture, woodturning, musical instruments, carving, boat-building
+ Straight, fine grain and attractive colour
– High degree of sapwood on each board

Pterocarpus dalbergiodes
Padauk

Origin Andaman Islands (Indian Ocean)
Sustainability Not yet endangered, though certified timber (lumber) unlikely
Used for Furniture, boat-building
+ Stunning colour, durable
– Interlocking grain difficult to work, blunts tools

Quercus alba
Oak, American white

Origin North America, Canada

Sustainability Not endangered, certified timber (lumber) widely available

Used for Joinery, furniture, construction work, flooring

+ Durable, strong, straight grain, inexpensive

– Grain lacks character

Quercus robur
Oak, European

Origin UK, Europe

Sustainability Certified timber (lumber) available, but not under threat

Used for Furniture, quality joinery, boat-building, timber (lumber) framing, veneer

+ Durable, hard and strong, distinctive grain and colour

– Natural defects mean wastage can be significant, expensive

Swietenia macrophylla
Mahogany, Brazilian

Origin South and Central America

Sustainability Endangered tropical wood, but certified timber (lumber) available

Used for Veneer, quality joinery, furniture, cabinetmaking

+ Crown-cut boards can produce flamed figure

– Grain can tear easily, prone to insect attack

Tectona grandis
Teak

Origin South-east Asia, West Africa

Sustainability Not endangered, though plantation-grown or certified timber (lumber) preferable

Used for Boat-building, garden furniture, decking, flooring

+ Water resistant and naturally durable

– Very expensive, blunts edge tools rapidly

Tilia vulgaris
Lime

Origin Europe

Sustainability Grows widely throughout Europe

Used for Woodcarving, toys, musical instruments, woodturning

+ Fine even grain cuts easily, not expensive

– Slight movement possible when dry, board ends can split

Ulmus hollandica
Elm

Origin Europe

Sustainability Increasingly rare, some certified timber (lumber)

Used for Chair seats, flooring, boat-building, veneer

+ Dramatic grain and colour variation

– Grain pattern makes it difficult to work, needs careful seasoning

project design

Design is an all too often overlooked area of woodworking, but taking the time to design before you begin can improve almost any woodworking project. This section covers the basic steps required to design a project from start to finish, although it is worth emphasizing the importance of going your own route and designing in a way that feels right to you. You can follow the steps in this chapter in a methodical fashion or combine just a few elements into a design process that works for you.

designing projects

It can be a daunting prospect to have to design something before you make it but designing is a simpler process than you might think. Breaking it down into several steps makes it much easier to understand, and so appreciate the benefits. All you are really doing when you design something is thinking through the practical and visual impact of an object, and then working out the various stages of its construction. Just remember: when designing, the right way is the one that works best for you.

Generating ideas

This first stage in the design process is often the most enjoyable. The fact that you are starting to plan a project means that you already have a basic idea of what it will be – for example, you may want to build a kitchen table or a corner cabinet. So now is the time to explore the possible options and gather inspiration from different sources to help you refine your ideas. Start by sourcing images in books and magazines; look for ideas that inspire you and are relevant to your project. Woodworking magazines are useful for sourcing ideas and plans but don't be afraid to spread your net wider while searching for inspiration. Many books and magazines on interior design and decoration specialize In providing visual stimulation of this sort and books on period furniture are a great resource too – even if you want your design to feature a modern aesthetic the chances are you can borrow some great ideas from history.

Use the Internet as well: take a note of the website of any furniture makers whose work you like for future reference; browse woodworking websites – the best ones often have gallery areas that showcase members' work, which can be a great resource in its own right. As time goes by and you work on more projects you will build up a valuable reference library of favourite books, magazines and websites that you can call on time and time again for inspiration – your library will become a more useful resource with each project you design and build.

Related info

Technical drawing (see page 154)
Material choices (see page 150)

Sketching

As you find inspiration from various sources, jot down ideas about how you can integrate these elements into your own project. Sketch ideas as you go – the key here is to create sketches that you can understand so don't worry about whether other people can read them. You will not have to frame your sketches and put them on public display, so just relax and get your ideas down while they are fresh so that you can review them later and combine the best ideas into a more complete concept. As long as you can understand your own sketches they are good enough. Sketching can help refine the style of the project and the choice of materials, and break quite complex ideas down into simple solutions.

You can sketch Ideas in two dimensions (2D) or three dimensions (3D). Both approaches are useful but you may find 2D sketching easier and particularly useful for woodworking projects – it does not require any perspective skills and can be drawn to scale easily, giving you a good idea of the visual proportion and finished size of a project. When drawing in 2D create separate views for the front, side and top of your project and apply approximate dimensions for reference to get a really good feel for how your finished project will look. Getting into the habit of drawing different 2D views will also make technical drawing easier if you decide to formalize your design with a finished orthographic drawing.

Notes on scale

Drawing to scale is extremely useful because it allows you to test the size and proportion of a design on paper. By scaling your measurements down to fit the size of paper you are working on it is possible to draw any design on virtually any paper size. Particularly useful scales are:

1:2 For small projects such as jewellery boxes.

1:5 Good for medium-size projects, such as wall and corner cabinets.

1:10 Best for larger designs such as tables and sideboards.

You can, of course, choose to work in any scale but dividing your measurements by factors of 2, 5 or 10 is particularly convenient. To use a scale ratio of 1:2 divide your finished measurement by a factor of 2. For example, a finished measurement of 100mm (4in) would be drawn 50mm (2in) long on paper. For scales 1:5 and 1:10 divide your finished measurement by 5 and 10 respectively to arrive at the drawn measurement.

material choices

Once you have a good idea of how the final project will look the next step is to plan the construction and decide how best to make it. However, before you can do this you need to establish the materials that will be used, because material choices will almost certainly dictate the best construction methods. There are two main categories of material to consider when woodworking – solid wood and manufactured board. Most projects can be made from either material, or a mix of the two, but both classes of material have very different characteristics that must be understood before establishing a method of construction.

Design principles for solid wood

Solid wood is a wonderful material to work with but it does provide the woodworker with a unique set of challenges:

- Since it is an organic material that grows naturally, no two pieces of solid wood are exactly the same, though the characteristics between two pieces of the same species will be very similar.
- Wood has grain that runs along its length. When wood is crosscut (cut across its width at right angles to its grain direction) end grain becomes visible. This can either be used for visual effect or hidden – as is the case with most traditional furniture.
- Unlike most manufactured boards, wood is a relatively unstable material. Even after it has been seasoned it will continue to move – especially across its width – because it continually acclimatizes to varying moisture levels in the air, which results in expansion and contraction. Even in modern centrally heated houses humidity levels will fluctuate, especially between seasons. For this reason, when working with solid wood, it is vital that your choice of construction method allows the wood to move without it cracking or distorting over time.
- Each species of hardwood and softwood will have different characteristics – see pages 140–145 for a list of common timber (lumber) types and their uses. Try working with as many different types of wood as possible as you progress from project to project. Some woods are easier to use than others – every woodworker will have their favourites and it is only through experimentation that you will find yours.

Related info

Wood characteristics (see page 124)
Buying wood (see page 132)
Construction methods (see pages 178–213)
Veneers (see page 134)
Acclimatising timber (lumber) (see page 162)
Wood and materials (see pages 120–145)
Hardware (see pages 266–279)

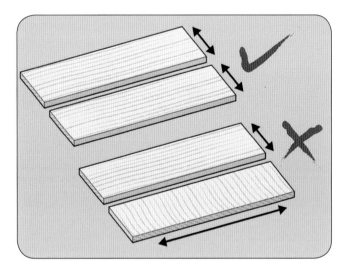

Jointed parts with opposing grain directions – such as mortise and tenon joints – are kept to minimum dimensions while long lengths should be joined with grain direction aligned.

Appropriate jointing methods

Traditional construction methods – such as mortise and tenon or dovetail joints – are used to connect solid wood components together. Most modern jointing techniques can also be used as long as the grain direction is considered carefully. The key to jointing solid wood successfully is in allowing each component adequate leeway to expand and contract without it affecting the overall structure. Frame and panel construction is a good example: The frame – made of square or rectangular section wood – is mortise and tenoned together. The frame is grooved on its inside face to take a floating solid wood panel, which is then free to expand and contract within the groove without exerting additional strain on the mortise and tenoned frame surrounding it – any movement across the width of the panel is isolated within the frame.

Another good example of a joint that allows wood to move is the dovetail joint, which is used to join end grain to end grain corners, such as in drawer and box construction. With dovetails the two jointed components should both have the same grain direction, so any expansion or contraction occurring over time is equalized between the parts. This is one reason why dovetails should never be made with opposing grain directions – end grain to long grain – between them.

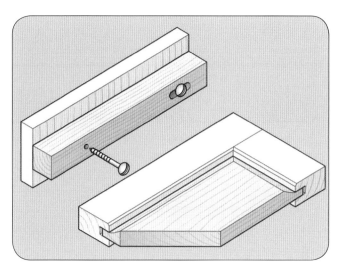

The slot hole in the upper example allows the two joined pieces room to move independently of each other In frame and panel construction (lower example) the panel is free to move within the frame without exerting pressure on the corner joints.

Allow for movement

When working out which construction method to use, remember that wood moves most across its width. This is why a panel requires a frame in order for it to move freely across its grain. If you make sure that the wood you use is always able to adjust slightly across its width, you should avoid problems – such as splitting – later.

Design principles for manufactured boards

Manufactured boards – such as MDF, chipboard and plywood – are manufactured to be as stable as possible so expansion and contraction is minimal with these materials. When movement does occur it is approximately equal in all directions, which minimizes the risk of additional stresses being introduced after jointing two or more pieces. Since manufactured boards are uniform in construction, building projects with these materials is a simpler process than with solid wood. With the right tools manufactured boards are easy to cut and joint, so projects will often be faster to complete than their solid wood counterparts. Although projects made with manufactured materials can tend to look cheaper and may lack a certain charm that is associated with solid wood, there are several things that can be done to disguise the material and improve its overall aesthetic.

Applying veneer Applying solid wood veneer to manufactured board is an excellent way to give the material a higher quality look and feel. Veneering requires its own set of skills and materials but is well worth the extra effort.

Pre-veneered board Pre-veneered manufactured boards can be purchased with a wide range of veneers. These boards will be much more expensive than the equivalent standard board material, but the results can be very impressive indeed from what is essentially an off-the-shelf product.

Solid wood edgings A solid wood edging is often used to protect the edges of a veneered board and enhance its overall look. If the edging is carefully matched to the surface veneer, the look of solid wood can be created giving the finished piece a high quality feel. Alternatively a contrasting wood can be used as edging to create visual interest

Mixing materials By using a combination of solid wood and manufactured boards, a balance can be struck to achieve greater economy and ease of construction, while maintaining a high quality appearance. Board material can be used for large flat surfaces, such as desktops and flat panels, while solid wood can be used for structural and decorative elements such as frames or mouldings. This approach can save both money and build time so mixing materials within the same design is widely adopted in commercial woodworking.

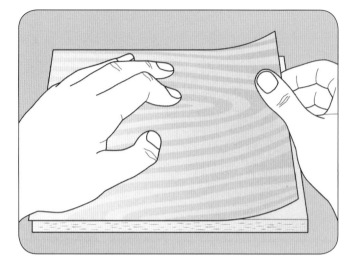

Applying veneer.

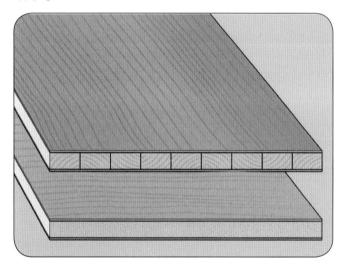

Pre-veneered board.

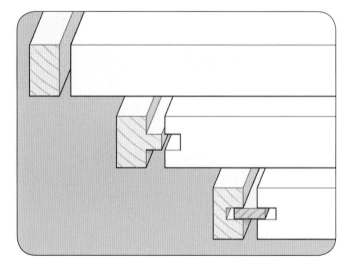

Solid wood edgings.

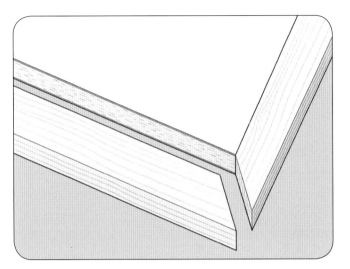

Pre-veneered board with solid wood edgings.

Appropriate jointing methods

It is generally not suitable to use traditional jointing methods – such as mortise and tenon or dovetail joints – to join manufactured boards because the inherent strength in the grain structure of solid wood is what gives these joints their durability. There are of course exceptions – a good example being plywood boxes or drawer sides and backs, which often feature finger joints (also called box lock joints). Since plywood is made up of thin layers of wood with the grain of each layer at right angles to that above and below, it lends itself well to this kind of joint. However, in most cases when joining manufactured boards other jointing methods must be used – for instance, biscuit jointing and dowelling both work very well and are quick to produce, especially biscuit joinery. Pocket hole screwing works well too, especially when wooden reinforcement battens are used on corner joints. There is also a whole host of knock-down (KD) fittings available for use with board materials.

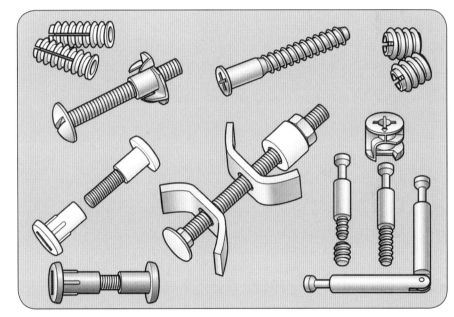

Knock-down fittings.

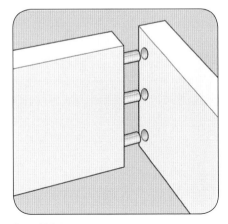

Dowel joint.

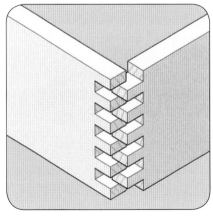

Finger or box joint.

technical drawing

Once you have decided on materials, the next step is to make a technical drawing – especially when designing and making furniture. Orthographic projection is a form of 2D drawing that incorporates three views – front, top and side. Additional views can be created – you might need to show both sides if a detail only appears on one – but for most projects it is not necessary.

Related info

Designing projects
(see page 148)

Producing an orthographic projection drawing

When an orthographic projection drawing is complete it should contain all the information required to make a design from start to finish. However, if you are both designing and making a project and the drawings are just for your own use, you can leave some elements unspecified until your project is partially built to get a feel for the finished piece before deciding on particular details.

- Secure a fresh sheet of layout paper to your drawing board.
- Work in pencil until your design is complete, then go over the pencil lines with a drawing pen if you wish.
- Use a solid 0.5 fibre tip pen for exterior lines and a 0.1 pen for interior lines. Hidden lines – showing components that would not be visible on a finished project – should be drawn as dashed lines.
- Use a T-square or parallel motion ruler to draw horizontal lines and a set square to draw vertical lines. Use scaled measurements suited to the paper size, making sure you use the same scale throughout the drawing. Mark any enlarged scale detail areas of the drawing clearly to avoid confusion.
- Start by creating a front view of your design in the top left quarter of the page (1st angle projection, see box right).
- Transfer horizontal lines with a T-square or parallel motion ruler from the front view to the right side of the page – these lines will form part of the side view. Next measure and draw in vertical lines using a set square to complete the side view.
- To create the top view (plan) in the bottom left quarter of the page start by drawing a horizontal line where you want the top view to start. Extend the left-most vertical line of the side view down until it intersects this horizontal line. Draw a 45-degree angle line through the intersection.
- Draw vertical lines from the side view down to the 45-degree line – where they intersect, a horizontal line is then drawn across to the left to form part of the top view. To complete the top view, draw vertical lines down from the initial front view at the top left of the page.
- Finish by adding dimensions, notes and details as required.

Orthographic angles

Angle projection refers to the layout of an orthographic projection drawing, which can be drawn at 1st angle or 3rd angle. It does not matter which method you choose but it's useful to be aware of which angle projection you are using.

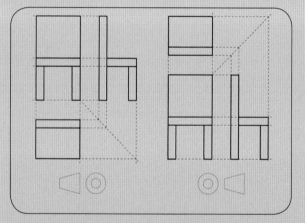

1st angle projection.   3rd angle projection.

Other views

You can also use the following views to explain complicated elements of your design:
Sections Cut-away views through a design to show methods of construction not normally visible.
Details Small elements magnified in scale are invaluable for working out joint measurements – for example, a 1:1 (full size) detail view of a joint can be used in the workshop to measure from.

Computer or drawing board

Technical drawings can be made on a drawing board or on a computer, and both methods have their advantages and disadvantages. Regardless of which you use, the principles of orthographic projection are the same.

Computer

+ Relatively easy to generate 3D drawings.
+ Multiple versions of the same design can be saved.
+ Requires no additional desk space if you already use a computer.
+ The undo button is always ready to save the day.

- Very difficult to use a physical ruler or a pair of dividers accurately on screen.
- Printouts are often not to scale and are limited to the capacity of your printer.
- With many CAD (computer aided design) packages it can be difficult to generate reliable 2D views from a 3D model.
- Available tools are often limited, especially with budget software, so the design may be compromised by the limits of the software.
- Bigger learning curve required to produce useful 2D orthographic projection drawings with a computer.

Drawing board

+ Drawings are handcrafted with curves often looking more natural.
+ Large paper sizes can be used – A3 or A2 sizes are ideal.
+ Using layout paper, sheets can be layered over each other to try different versions of the same design, which also record the design progression.
+ Scale is constant so rulers and dividers can be used directly on the drawing with accurate results.
+ The finished drawing can be taken to the workshop and notes added as the project progresses.
+ A drawing board and T-square can easily be made in the workshop.

- Additional desk space is required to house a drawing board.
- There is no undo button.
- Subsequent versions of a design have to be traced again by hand.

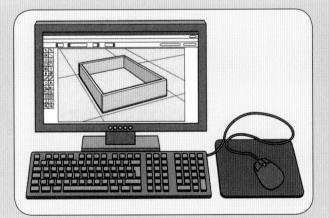

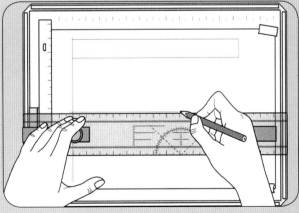

workshop rods and templates

As well as sketches and technical drawings of your project, you may need workshop rods and/or templates during the construction process.

Workshop rods are full size 2D drawings, which provide a highly accurate 1:1 (full-size) scale reference that can be used at the bench to take measurements. They are usually created on hardboard because of its light weight and cheapness, but most smooth-surfaced board materials would do equally well. Workshop rods can be particularly useful aids when a project contains curvature or complex joinery, because the work can be placed directly on the rod to check how it is progressing at any time during the construction process.

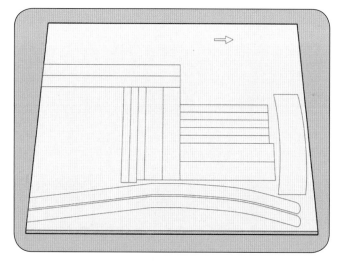

> **Related info**
>
> Technical drawing (see page 154)

Templates are often made for curved or complex details. They are drawn full size before being cut out and accurately shaped by hand. They are particularly useful for repeat work because the shape of a template can be transferred to a workpiece by drawing around its perimeter. For batch runs (production of more than one identical component at a time) templates can be used with bearing-guided router cutters to cut profiles directly from fresh material without any need for additional marking.

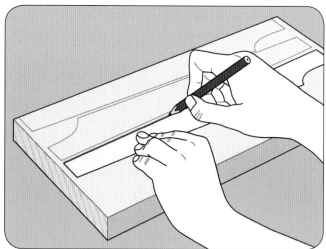

Using off-cuts

Birch plywood and MDF off-cuts can be excellent for making templates. Keep a small box full of useful size off-cuts for jig and template making.

cutting lists

The cutting list is the backbone to any planned project – it is used to record all the components once the design is complete. Having a cutting list to hand before starting a project makes the process of preparing your material much easier and quicker – without one you will soon lose track of the various components that need to be made.

Making a cutting list

A cutting list should be drawn up with the following columns:

Description This describes the component being specified.
Length The finished length measurement.
Width The finished width measurement.
Thickness The finished component thickness.
Number off The number of components of each size that need to be made.
Notes Any peculiarities or special requirements for each component can be noted here.

Related info
Technical drawing (see page 154)
Buying wood (see page 132)
Sheet size (see page 139)
Planning and costing (see page 159)

A cutting list template is given on page 158. To create your cutting list, first start by working out how many components will be needed to complete your project. Give each component a name. Duplicate components, such as table legs, should share the same name. List each component in the description column, and then note the number of each required in the number off column. Next, add any important points in the notes column before filling in the length, width and thickness columns to complete your cutting list.

Cutting list

Project title.. Date ....

Description	Length	Width	Thickness	Number off	Notes

planning and costing

Having worked out a cutting list, the last phase of the design process is to plan the various stages of the build. It is useful to write this down as a series of bullet points so that you can check off each stage as you progress through the project; creating a simple list like this can be a huge time saver, especially if you have a small workshop where individual machines and tools have to be set up to complete each phase of a project. It will also ensure you do not forget anything fundamental along the way.

Working out the cost of a project prior to starting manufacture can make or break it. By creating a cutting list you have done most of the hard work already – use it to work out how much material you will need to complete the project. Don't forget to include fittings and fixtures at this stage too.

If you are using PAR (planed all round) wood – commonly available in softwood form from builders' merchants and DIY stores – then the amount of material you need can be worked out in linear lengths because width and thickness measurements are standardized and the surfaces have been planed square. Manufactured boards are also sold in standard sizes with thickness being the main variable. For details of common sizes and thicknesses see page 139.

Buying sawn wood is a little trickier because you need to work in either cubic feet or cubic metres (see page 132). Even if you prefer to work in metric it is advisable to work out your cutting list in cubic feet, simply because one cubic foot can be visualized as a plank 1in thick by 12in wide by 12ft long. A cubic metre, on the other hand, is such a large amount of material that it is hard to visualize for domestic projects. Sawn wood is usually bought from specialist timber (lumber) merchants – some offer just a small range of native hardwoods, while others offer a huge variety of different species. Prices can vary widely from merchant to merchant, so it is worth hunting around for a good supplier that you can trust and then build a relationship over time. When sourcing sawn wood ask for prices in cubic feet, and then work out approximately how many cubic feet you will need and factor in an additional 15–20 per cent wastage.

Related info

Technical drawing (see page 154)
Cutting lists (see page 157)
Buying wood (see page 132)
Manufactured boards (see page 136)
Hardware (see pages 266–279)

Checking timber (lumber)

Always check that PAR timber (lumber) is straight and square prior to use – because of the way this. type of wood is stored it is very common for it to be out of true by the time it reaches its end user.

material preparation and basic techniques

The first stage on any project using rough-sawn timber (lumber) – most hardwood is supplied rough-sawn – involves preparing the material for use. Wood must be acclimatized before squaring, thicknessing and then cutting to length accurately, either with hand tools or machinery. These are all fundamental skills for working with wood and achieving a successful project, so taking the time to get these basic stages right will pay dividends as you progress and take on more challenging woodworking projects.

acclimatizing timber (lumber)

Although the process of seasoning wood brings its moisture content down to a usable level, even after it has been seasoned it continues to adjust to moisture levels in the surrounding air. If you don't plan for this in advance further shrinkage or distortion is likely to cause problems.

The acclimatizing process

Acclimatizing involves placing already seasoned wood in a room that has a similar temperature and humidity level to the room that it will occupy as a finished project. It is not advisable to use unseasoned (green) wood for indoor projects, because as it dries the subsequent distortion will be severe. Nearly all commercially available wood sold through timber (lumber) merchants is kiln-dried, so wood from a reputable supplier will usually have been seasoned to a moisture content level of between 14 and 18 per cent. Placing this wood indoors can bring the moisture content down further over a series of weeks, to a level more suited to domestic projects. If acclimatizing is carried out before squaring up and cutting to finished length ready for use, it minimizes the risk of wood moving significantly during and after you build a project.

Correct moisture content levels

15–16 per cent is suitable for partially exposed joinery such as doors and windows. It is also a suitable level for outdoor furniture.

9–13 per cent is ideal for domestic furniture, with 9 per cent better for a modern continually heated property with double glazing and 13 per cent more suited to an older property with less efficient heating and minimal draught exclusion.

7–8 per cent is better for areas that are exposed to a lot of dry heat, such as airing cupboards.

The important factor is to allow the wood to rest for a period of time in an appropriately heated room to allow it to adjust to the conditions. If it is out of sight during this phase then all

the better, so a space behind a sofa or under a bed is ideal. If you are lucky enough to have a centrally heated workshop, or a workshop that is part of your house, then storing wood in your workshop could well provide the acclimatization necessary. When stacking wood to acclimatize it indoors use battens approximately 25mm (1in) thick to create spaces between the stacked layers. This ensures consistent airflow and allows all the wood to adjust at the same rate, which in turn minimizes the risk of distortion. After the acclimatization process the next stage is to sort the wood ready for planing – this stage is called setting out.

Related info

Seasoning (see page 130)
Workshop storage (see page 18)
Caring or stored wood (see page 133)

Acclimatizing timescales

Softwood Allow to acclimatize for two weeks per 25mm (1in) of thickness.
Hardwood Allow to acclimatize for four weeks per 25mm (1in) of thickness.
If you don't have the time to allow your wood to acclimatize fully, then even a short period is better than no time at all. Leaving wood indoors for as little as a week prior to using it is better than taking it straight into your workshop from the cold, and often damp, conditions found in many timber (lumber) yards.

The timescales given above are only a rough guide, because the correct acclimatization time will vary from species to species – a very dense hardwood will take longer to acclimatize than a relatively lightweight one, for instance. For accurate readings use a digital moisture meter; a good meter will give you moisture level readings to one percentage point.

Meters

When buying a moisture meter choose one from a brand with a proven track record and moisture conversion tables based on Timber Research and Development Association (TRADA) data.

setting out

Setting out is the process of organizing wood, marking it and cutting it to approximate length prior to planing it square. In order for this stage to go well you will need your cutting list – treat it like a check list, ticking off each individual component as you allocate material.

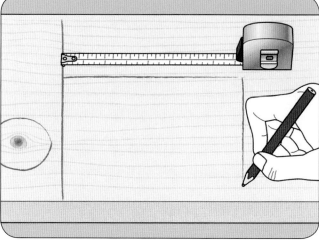

1 Inspect the wood for faults and use a contrasting colour piece of chalk or a medium soft pencil to highlight areas of concern to be avoided, such as cracks, splits and knots.

2 If setting out rough-sawn wood, draw approximate boundary edges in pencil on each sawn board as a reminder of which components it will produce.

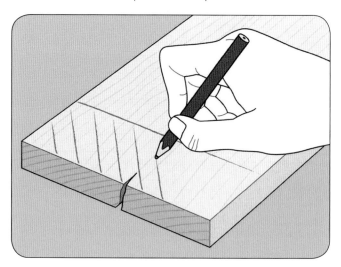

Related info

Cutting lists (see page 157)
Defects and blemishes (see page 126)
Measuring and marking out (see page 33)

3 Inspect all end grain for defects and hidden splits. If you find any, draw a line across the grain well clear of the defect and crosshatch the waste area as a reminder that this portion of the wood should not be used.

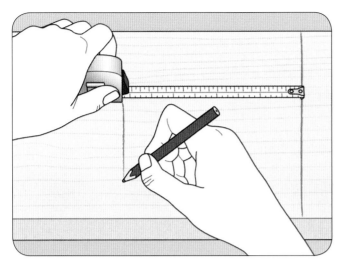

4 Measure each piece of wood to approximate length, leaving enough spare material at either end to cater for jointing plus a little wastage – this allows a secondary clean crosscut to be made at each end after the wood has been planed and thicknessed. Particularly short components can be left as part of longer boards at this stage, but remember to increase the wastage accordingly. The objective at this point is to set out the material needed to complete your cutting list and end up with a stack of wood with each piece manageable in size.

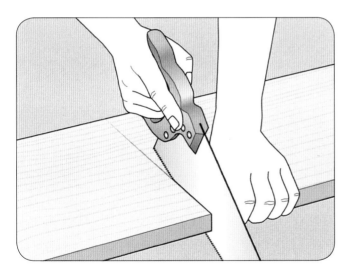

5 Having double-checked all measurements and marked each piece for its intended purpose, use a pair of trestles and a handsaw to cut the material to approximate length.

planing primary surfaces

If you are using rough-sawn boards (wood in its pre-planed state) supplied by timber (lumber) merchants, the sawn faces will need planing square prior to use. Both hardwood and softwood can be bought rough-sawn. PAR timber (lumber) should also always be checked before use, because it is not uncommon to find that it is out of square by the time it reaches the end user. If this is the case then the following techniques can be used to true it back up accurately.

Wood should be squared up in the following order: face side; face edge; width; thickness. Once these four operations have been completed, the ends can be crosscut square and to final length.

Related info

Planes (see page 44)
Power planers (see page 76)
Planers and thicknessers (see page 108)
Universal machines (see page 118)
Safety (see page 92)

Face side

The face side is the first side to be planed. This can be done either with a bench plane or by using a surface planer (jointer) or – as is common in Europe – a combination planer thicknesser (jointer planer) that planes both the face side and face edge surfaces and then converts to a thicknesser (thickness planer) to bring the wood down to a specified thickness. For best results, plane with the grain direction to avoid tear-out. The objective when planing the face side is to achieve a totally flat surface that will serve as a reference surface to which all subsequent surfaces can be squared.

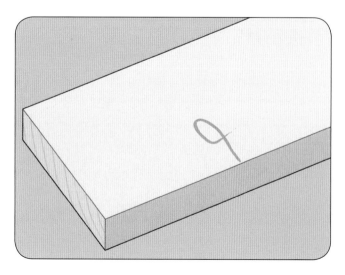

Planing the face side by hand

To plane the face side by hand start with a coarsely set jack plane to remove lumps and bumps.

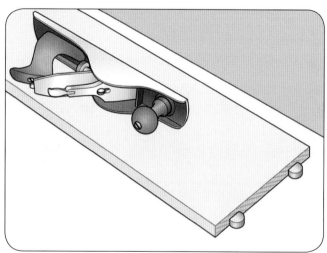

1 Secure the board to a good, solid bench top, choosing the side of the board most easily supported to lie flat on the bench. An end vice and bench dog setup is ideal for securing boards like this, but a simple workshop-made end stop will suffice if you don't have a dog system on your bench.

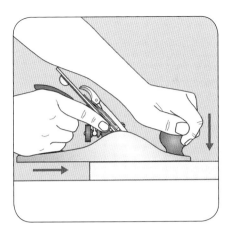 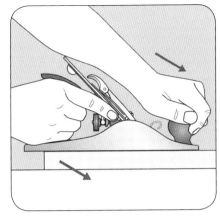 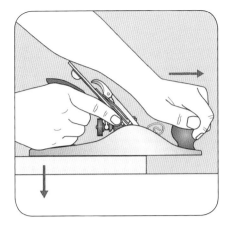

2 Start by applying pressure at the front of the plane.

As you progress the plane towards the middle of the board, transfer pressure to the middle of the plane.

Transfer pressure to the rear handle as you come to the end of the cut. As you develop this technique you will find it easier to achieve the variable pressure process in one fluid movement – with practice you will not even have to think about it.

material preparation and basic techniques

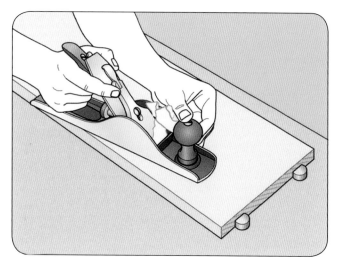

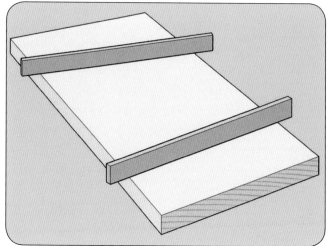

3 When the material feels relatively flat, set the plane to take thinner shavings. If possible, switch to a longer fore plane at this point because the extra length will make a flatter surface easier to achieve. If you don't have a fore plane, good results are still achievable with a jack plane – you will need to take a little extra care, but this will go towards developing valuable hand skills that no amount of money can buy.

4 Use a pair of winding sticks to sight along the length to check for twist. Winding sticks exaggerate any twist present, so you can assess how flat your workpiece is. Mark high points, and then remove them with a plane before rechecking. Repeat until your face side is completely flat and true. The final step is to mark with the face side mark, which – as you will see later – is an invaluable reference when thicknessing.

Planing the face side by machine

As with hand planing, for best results plane your wood with the direction of the grain to avoid unsightly tear-out.

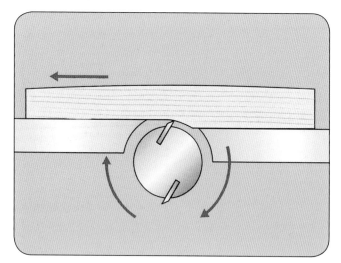

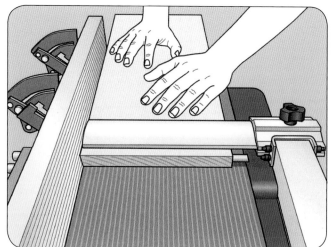

1 Sight down the board to establish any bowing or twisting. If the board is bowed then the concave side should be planed first, because it will have more points in contact with the infeed and outfeed tables than a convex curved side will – this generally results in more accurate cuts with fewer passes over the machine.

2 Whether you use a dedicated surface planer (jointer) or a planer thicknesser (jointer planer), the process is the same. Take great care because these machines are one of the most common causes of workshop accidents – note the safety information on page 92. Always keep your hands away from the cutterblock as you feed wood over it, even with it fully guarded.

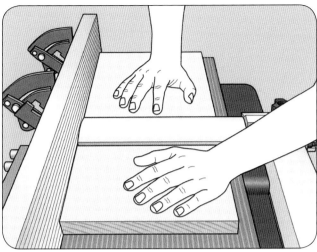

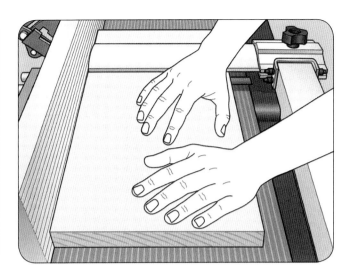

3 With the machine running, start by applying pressure on top of the wood as you feed it over the infeed table; keep the pressure steady and feed the wood at a gentle but constant rate. As the wood passes over the revolving cutterblock wait for a safe amount of wood to feed over the outfeed table, then lift and transfer your left hand over to the outfeed table side, ensuring that it is well away from the cutterblock. Continue applying gentle pressure while keeping the feed rate as constant as possible.

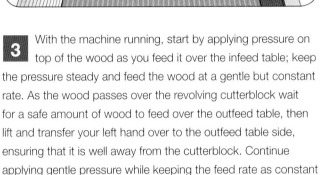

4 As the wood progresses over the cutterblock there will come a point when you need to lift and transfer your other hand from the infeed to the outfeed side to continue feeding the wood over the cutterblock safely.

Using a pushblock

Inexperienced users should always use a pair of push blocks to feed the wood over the cutterblock instead of their bare hands. This adds a physical barrier between the revolving cutterblock and the user, but care still needs to be taken when using a surface planer (jointer). A push block should also be used when planing short pieces or thin timber (lumber), as there is a tendency for the material to lift as it passes over the knives. Always slide the bridge guard across to the edge of the pushblock.

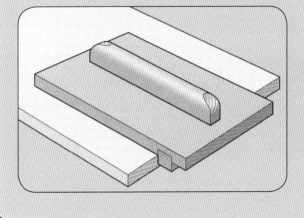

5 After each pass turn the wood over to examine its surface before running it over the planer (jointer) again. If you are planing rough-sawn boards then you should see areas of smooth planed wood and contrasting areas of unplaned sawn wood; continue passing the wood over the machine until it is completely flat and smooth. Check for tear-out regularly – if necessary try rotating the wood before feeding it over the machine again, and then compare results. When you are satisfied that the entire surface is planed flat, add the face side mark with a pencil.

Face edge

The face edge is planed using the face side as a reference –
it is planed at right angles (90 degrees) to the face side
and is the second reference face.

Planing the face edge by hand

Planing the face edge by hand is achieved with a No.5 or a
No.5½ jack plane, a No.6 fore plane, or even a longer No.7 try
plane which is ideal for planing long edges. If the board is narrow
and relatively short it can be placed in a bench vice. Longer
boards can still be supported in a vice, but additional support
will be required at one end.

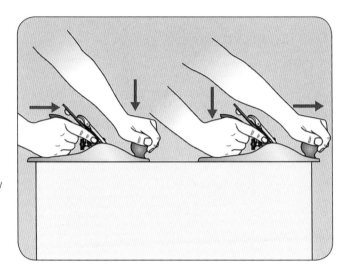

1 Use the same planing technique as when planing the face
side, starting with pressure at the front of the plane, then
changing to the middle as the cut is progressed and ending with
pressure at the rear of the plane as it exits the cut. Keep the cut
as close to a right angle with the face side as possible.

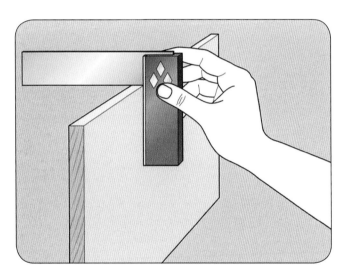

2 When you achieve a cleanly planed face edge surface,
take a square and check the surface is at right angles to
the face side at various points along its length. Hold the board
towards a light source as you check, because this will highlight
any discrepancy. Where the surface is not square, lightly mark it
with a pencil and take shallow cuts with a plane before checking
again with the square. Repeat this process until the face edge is
completely square to the face side along its entire length.

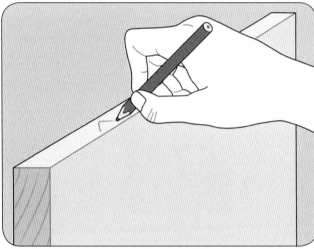

3 The final step is to add the face edge mark with a pencil.
You now have the two most important reference surfaces
complete, from which the finished width and thickness of the
board can be produced.

Planing the face edge by machine

Creating a face edge with an overhand planer (jointer) is relatively straightforward, and uses the same techniques as when planing the initial face side. Make sure hands and guards are in the proper position while feeding the material at a constant rate; this time the side fence is used.

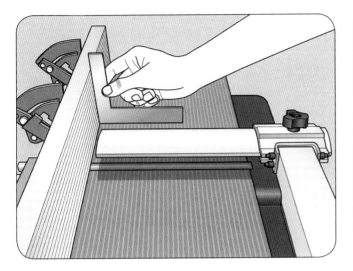

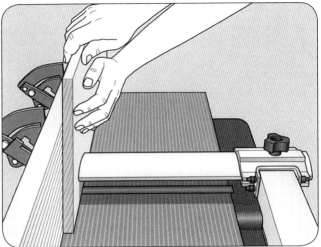

1 Before beginning check the fence is at right angles to the infeed table with a square, and then lock it securely in place. Adjust the guards to suit the timber (lumber) section you intend to plane. With the machine running, place the face side against the fence on the infeed table, with the edge that will be the face edge flat on the infeed table.

2 When you are satisfied that everything is correct feed the material over the cutterblock, maintaining light pressure to keep the face side flat to the fence. Inspect the edge after each cut – if the fence is set accurately to the table there should be no need to check your edge with a square. When complete, add the face edge mark with a pencil next to the face side mark.

Using a planer thicknesser (jointer planer)

Set up your planer thicknesser (jointer planer) knives accurately – with a bit of practice this can be done quickly and will save you a good deal of frustration when using the equipment. Refer to the manufacturer's instructions for your machine for the most appropriate method of achieving this.

Always mark the face side and the face edge. If you don't you could thickness the wrong side and end up with out of square stock.

Guards are there for a reason. Improper use of the planer thicknesser (jointer planer) is one of the most common causes of workshop accidents. Always make sure your guards are correctly set before switching the machine on.

Check your grain direction before you begin. If you are not sure which way to plane, try both directions then mark your material with an arrow when you have established which direction gives the best results.

cutting to width

There are many ways to cut wood to width. Most common methods involve using a rip fence on a table saw or bandsaw but you can also rip – cut with the direction of the grain – wood with a handsaw.

Ripping

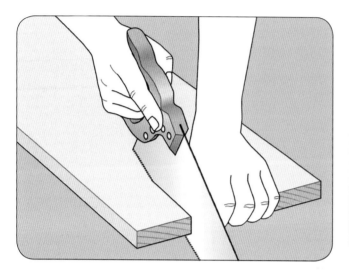

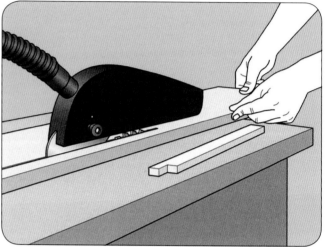

1 For best performance, it is worth considering dedicated ripping and crosscutting handsaws or blades, which are optimized for one type of cut or the other and will therefore cut more efficiently.

2 Most general purpose circular saw and bandsaw blades – as well as hard point handsaws – are suitable for both ripping and crosscutting. Investing in general-purpose blades makes good sense for most domestic workshops.

Related info

Bandsaws (see page 94)

Table saws (see page 98)

Cordless power tools (see page 62)

Portable saws (see page 68)

Saws (see page 40)

Safety (see page 92)

Working in stages

Always take several shallow cuts when working with the planer (jointer) or thicknesser (thickness planer), rather than making one deep cut.

thicknessing wood

As with almost all woodworking tasks, thicknessing can be done either by hand or by machine.

Using the face side mark

When thicknessing, the face side mark is always used as the reference face – the face side has already been planed flat so its reverse surface is the one that is thicknessed parallel.

Related info

Planes (see page 44)
Power planers (see page 76)
Planers and thicknessers (see page 108)
Table saws (see page 98)
Bandsaws (see page 94)
Handsaws (see page 40)
Universal machines (see page 118)
Safety (see page 92)

Thicknessing by hand

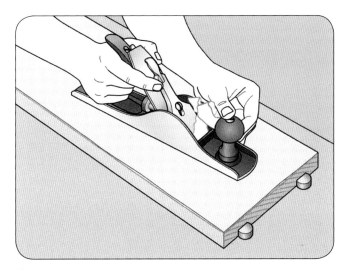

When thicknessing by hand, first mark the desired depth around all four edges using a marking gauge. Use a bench plane to plane away material; when the depth mark is reached an even thickness has been achieved. Ideally, start with a scrub plane for rapid stock removal, followed by a jack or fore plane to finish.

Thicknessing by machine

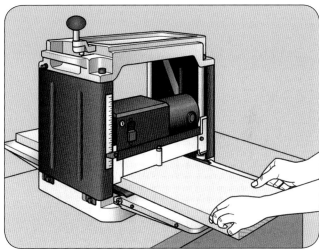

Thicknessing by machine is relatively easy, using either a planer thicknesser (jointer planer) in thicknessing mode or a dedicated thicknessing machine (thickness planer). Bring the material down to the desired thickness in small increments.

cutting to length

Crosscutting to length – cutting at right angles to the grain direction – is another fundamental woodworking skill, which will be used numerous times on any project involving solid wood.

Crosscutting by hand

A bench hook is essential for accurate crosscutting by hand. This simple bench-top holding device is made from hardwood and can either be bought or made in the workshop.

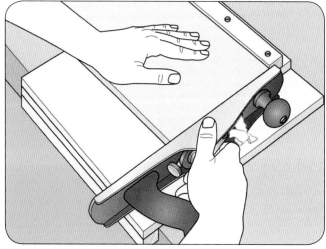

1 The bench hook supports your wood during cutting, while protecting the bench top from accidental saw cuts. Use a back saw in combination with a bench hook for accurate crosscutting of small- to medium-size components.

2 Place the workpiece on a shooting board and use a bench plane on its side to skim thin shavings of end grain to square and neaten the crosscut.

Squaring up

If you decide to make either a bench hook or a shooting board, it is vital that all components are as square as possible – discrepancies can lead to inaccuracies when using them, especially with the shooting board.

Related info

Workbench basics (see page 16)
Saws (see page 40)
Cordless power tools (see page 62)
Portable saws (see page 68)
Woodworking machines (see pages 90–119)
Safer woodwork (see page 22)

Crosscutting by machine

Cutting accurately to length can be achieved on a variety of machines, including the table saw, mitre saw and radial arm saw. The mitre saw is particularly popular because of its combination of accuracy, small footprint and low cost.

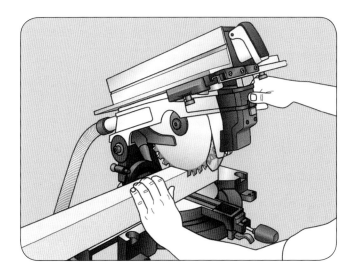

cutting manufactured boards

Hand tools are not well suited for cutting manufactured boards, although a hardpoint saw can be used to rough-cut them to size. These boards don't generally plane well because of their structure; sophisticated bonding resins are used in their manufacture, which tend to dull hand tool edges quickly.

Cutting methods

A good table saw fitted with a TCT blade and rip fence will give good results, but only within the size constraints of the saw. Most full-size sheets of manufactured board are too large to be cut down on a table saw in a small- to medium-size workshop.

Related info

Power tools (see pages 60–89)
Woodworking machines (see pages 90–119)
Saws (see page 40)
Cordless power tools (see page 62)
Routers and cutters (see page 79)
Dust control (see page 20)
Safer woodwork (see page 22)

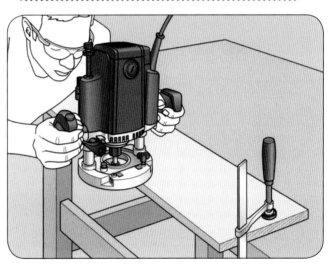

1 One of the best and most accurate methods of converting manufactured boards to finished size is by using a plunge saw or circular saw with a guide rail. This setup can be easily manoeuvred to the material, making it ideal to use on an existing bench or work surface, even outside in good weather.

2 Jigsaws and circular saws without guide rails can also be used, but the cut edges will need to be squared up afterwards if any degree of accuracy is required. This can be achieved with a router fitted with a bearing-guided profile cutter, which is run along a straightedge to trim the cut straight and square. This is a great approach if you are woodworking on a budget and only have a few power tools.

construction methods

Good jointing technique is crucial to achieve high-quality results in woodworking. There are a multitude of joints to choose from, each with its own distinct advantages and disadvantages, and part of the skill in producing joints is in choosing the correct one for the job at hand. This chapter will guide you through choosing and cutting a wide range of hand- and machine-made joints, for a variety of different woodworking tasks.

choosing the right joint

The process of joining pieces of wood together to form solid structures has evolved from generation to generation, to the point where little improvement can now be made. The methods of jointing employed today – especially those that are hand-cut – have been tried and tested to the point of exhaustion, along with every possible variation, to establish a vast array of possibilities and well proven techniques. Any innovation that does occur tends to be in the adaptation of traditional methods to modern machine tools, many of which are more relevant to production lines and the use of manufactured boards. In the small domestic workshop, hand-tooled techniques are still widely used and in many cases have a visual edge over their machine-cut counterparts.

Types of joint

Edge joints Also known as butt joints, with these the edges are butted or rubbed against each other and then glued. Edge joints can be formed with or without reinforcement.
Interlocking joints Male parts are cut to fit corresponding female components, maximizing gluing area and strength.
Mechanical joints These secure parts together while allowing controlled movement to occur.
Knock-down (KD) joints Usually made of metal or a tough plastic such as nylon, KD fittings have been engineered for the production of most types of furniture. Because KD fittings are essentially off-the-shelf solutions to jointing rather than joints made from scratch, they are not covered in this chapter – for more information on KD fittings refer to the Hardware chapter, pages 273–275.

It is worth noting that some joints will fall between two categories – a reinforced edge joint being a good example, because the reinforcement transforms an edge joint into an interlocking joint.

Skills

There is a good deal of skill involved in marking out and cutting solid wood joints by hand. However, time spent practising their production will be well spent, because each time a joint is produced satisfactorily the confidence and enjoyment of the maker increases too. For most joints there are hand- and machine-cut options and of course it is up to you to decide which is the best method to use; in this chapter you will find a good balance of the two different approaches.

Related info

Wood characteristics (see page 124)
Design principles for manufactured boards
(see page 152)
Design principles for solid wood (see page 150)
Hardware (see pages 266–279)

Choosing a saw

If you choose the hand-cut option, you will need the correct saw to cut the joint to achieve a good result. Depending on the job at hand, this will either be a dedicated crosscut saw or a rip (tenon) saw – for some tasks, a tenon saw can also often be used as a substitute for the crosscut saw. You may also need a dovetail saw, which is a smaller version of the tenon saw with finer teeth.

marking out

The first stage in creating any joint is marking out and unless you are using a fully mechanized machine process this is still the exclusive job of hand tools. Good marking habits will increase the accuracy of your work while also minimizing mistakes.

Related info

Measuring and marking (see page 26)
Measuring and marking out (see page 33)

To mark out the joints in this chapter you will need the following tools:
Try square (see page 26) or combination square (see page 34)
Marking gauge (see page 27)
Mortise gauge (see page 30)
Marking knife (see page 27)
Sliding bevel (see page 26)
Tape measure (see page 26)
Steel rule (see page 26)
Dividers (optional) (see page 27)
Sharp pencil – 2H is ideal (see page 26)

edge, butt and mitre joints

Edge joints are made when two edge surfaces are brought together and glued. If joined long grain to long grain so the grain is running parallel on both components, the pieces can shrink and expand at the same rate across their respective glue line. With good adhesive long grain to long grain edge joints are extremely strong and in most cases require no additional reinforcement.

Edge joints

This method of edge jointing is commonly used to glue wide panels together – see also rub joints, right. The strongest edge joint is long grain to long grain (A–C). Edge joints running end grain to end grain (A–B) are very weak, not only because end grain soaks glue away from the surface like a sponge but also because the wood fibres are at right angles to the glue surface. End grain to long grain edge joints (B–D) are also quite weak and suffer from the added risk of differing shrinkage rates between components, which inevitably puts added stress on any glued area.

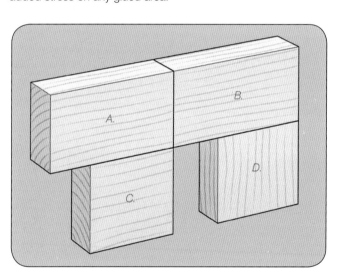

Rub joints

Rub jointing is a method of joining boards edge to edge along their length, with the grain between boards running parallel. Rub jointing is ideal for creating table tops, panels or any other large area that needs a flat panel of solid wood wider than the board width available.

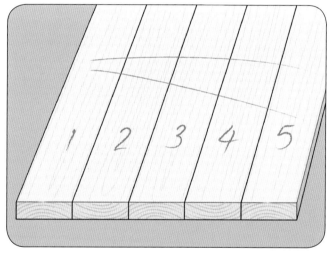

1 After preparing the wood to the correct width and thickness, lay the boards to be jointed on the bench and arrange them in order. For the most stable construction alternate the annual rings so that they cancel each other out should the individual boards cup over time. This may have a visual impact on the overall look of the panel, and if it will compromise the grain pattern and colouring then careful consideration should be given as to whether stability or visual effect is more important. Next mark each set of corresponding edges with a series of marks so their position in relation to each other is recorded – you can either number the boards or use diverging lines, as shown above.

> **Related info**
>
> Choosing the right joint (see page 180)
> Marking out (see page 181)
> Wood characteristics (see page 124)
> Adhesives and assembly (see pages 214–223)

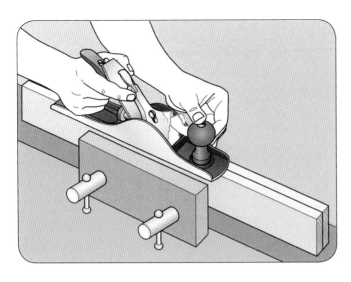

2 Shoot the boards with a plane to ensure a good match; start by securing adjacent boards together in a bench vice with their corresponding edges lined up. Take a long plane – ideally a fore or try plane, but with care a jack plane will also do the job – and take thin shavings along the edges of both boards. When continuous shavings are achieved along the entirety of both board edges it is a good sign that they are well aligned.

3 Put the two boards together on the bench top and test the fit by looking for gaps; ideally you should have a good tight fit. A slight concave curve leaving a very small gap towards the middle of the joint can be an advantage, because when it is cramped (clamped) together the middle will pull tight, minimizing the risk of separation over time at the ends of the joint. However, if you cannot remove the gap with hand pressure alone, then it is too large and the boards will have to be planed again. If the curve is slightly convex so there are gaps at either end of the join, then the boards will also need to be replaned because the chances of the join separating at either end later are high.

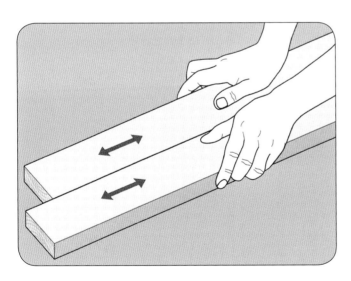

4 A good straight rub joint can be glued without any cramps (clamps). The joint's name is derived from the action of rubbing the two edges together after glue has been applied, which creates a vacuum that holds the pieces together while the glue sets. As a general rule this approach works well on small pieces, but for larger work it is normal practice to use cramps (clamps). Place the cramps (clamps) alternately above and below the workpiece to equalize the pressure and ensure the boards do not become misaligned during assembly.

Reinforcing edge joints

Rub joints can be reinforced if strength is a concern, although it should be noted that with modern adhesives there is little need for reinforcement under most circumstances. However, adding an interlocking feature – such as a biscuit or loose tongue – will aid assembly because the interlocking element will stop the boards sliding around under pressure during glue-up.

construction methods

Butt joints

Butt joints can be used for basic box and drawer construction, although some form of reinforcement is necessary in order to give them adequate strength. They can be used for both wood and manufactured boards. Accurate crosscutting is essential because the quality of cut will be visible on the finished joint – if crosscutting by hand then a bench hook and shooting board should be used for maximum accuracy.

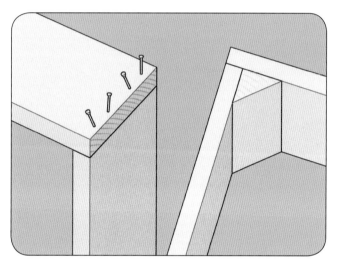

Reinforcement options for butt joints

Biscuits These can be used to invisibly strengthen butt joints. This method is particularly useful when joining MDF or plywood – see Biscuit jointing on pages 196–197.

Dowels Either blind or through dowels work well, particularly with solid wood – see Dowel jointing on pages 197–199. If inserted all the way through the exposed end through dowels can form an attractive feature.

Countersunk screws The heads can be concealed with solid wood plugs.

Pocket hole screws These can be driven in from behind the joint, but should only be used if the rear of the joint will be concealed.

Pins Driven in at opposing angles, these offer a quick and effective solution but can cheapen the look of the end result.

Mitre joints

The mitre joint is desirable because of its aesthetic properties; it offers a continuous flow of grain from one component to the next, particularly when cut from the same length of wood. The mitre joint is essentially an angled butt joint, and like the butt joint it will need some form of reinforcement to give it structural integrity if it is to be used under any kind of load.

Cutting mitre joints by hand

Mitre joints are ideal for frames, boxes and even some forms of cabinet construction.

1 If cutting a 45-degree mitre, then first mark the mitre with a marking knife and mitre square (or combination square). If a different angle is required, then use a protractor to set a sliding bevel at the desired angle. Mark the angles on both sides of the cut using a square to join them up so that all four faces are marked.

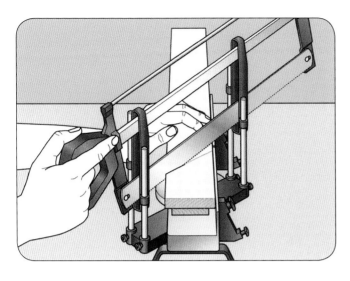

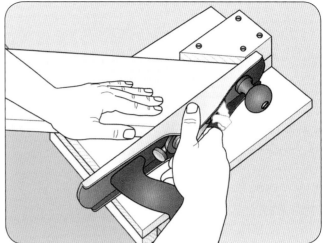

2 Cut the waste material away, using either a back saw with a mitre box or a dedicated mitre saw.

3 Use a mitre shooting board and a sharp plane to trim the surfaces accurately. Alternatively, the mitre can be secured in a vice, with a piece of waste material supporting the back edge to prevent any tear-out, and then carefully trimmed with a block plane.

Cutting mitre joints by machine

There are several machines capable of cutting accurate mitres; the table saw, radial arm saw and dedicated sliding compound mitre saw (SCMS) are good choices. Of the three, the SCMS is probably the most accessible and cost effective option. Many are now fitted with lasers, making the job of cutting mitres or compound mitres even easier.

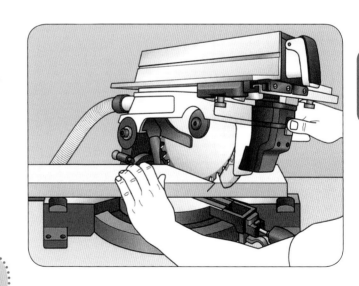

Accurate mitres

For accurate mitres, cut them slightly oversize and then trim them down on a dedicated mitre shooting board.

Reinforcing mitre joints

Splines Pieces of veneer or thin sections of wood are glued into slots cut into the joint. The advantage of veneer splines is that a single cut with a dovetail or tenon saw often provides just the right width for the veneer to be inserted, making this a relatively quick technique. This method is particularly useful for box construction because of its decorative appearance.

Dovetail keys Although more complex to produce, these yield very strong results and have the added bonus of being quite decorative. This technique works very well with box and small cabinet construction.

Loose tongues These provide a very strong joint and can be made from solid wood or manufactured boards such as plywood or MDF. Care should be taken not to expose the ends of the loose tongue in visible areas.

Dowels, biscuits and Festool's Domino system All of these can be used to strengthen mitres – particularly in cabinet construction where structural strength is important.

Wedges For picture framing where the back of a frame is not visible, the use of wedges or staples driven in to join two mitred components together will significantly strengthen the joint.

lap and scarf joints

Lap joints are used to join the corners of boxes, drawers and cabinets. Scarf joints are used when two pieces of wood need to be joined end to end (end grain to end grain).

Lap joints

With a lap joint, a rebate (rabbet) is cut on one side allowing the other side to fit snugly into place without its end grain showing. The small amount of end grain visible at the end of the rebate (rabbet) is referred to as the lap. Mitred lap joints offer a more aesthetically pleasing option because no end grain is visible but they are more difficult to cut.

Related info

Choosing the right joint (see page 180)
Marking out (see page 181)
Wood characteristics (see page 124)
Adhesives and assembly (see pages 214–223)

Cutting lap joints by hand

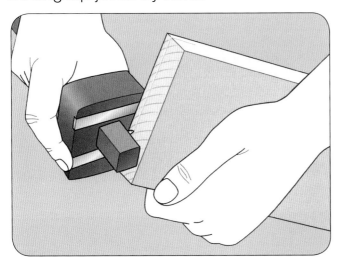

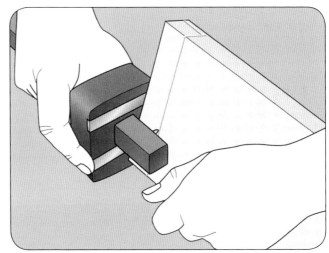

1 Take a marking gauge and set it to one quarter of the thickness of the wood, then mark the lap on the end grain from the outside face of the piece to be rebated (rabbeted). Also mark both sides of the lap.

2 Take a second marking gauge and set it to the thickness of the piece that will sit in the rebate (rabbet) behind the lap, and then mark the other side of the rebate (rabbet). Bring the two sets of lines together with a square, then crosshatch the waste area with a pencil. Place the workpiece upright in a vice for the next step.

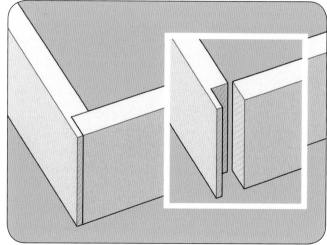

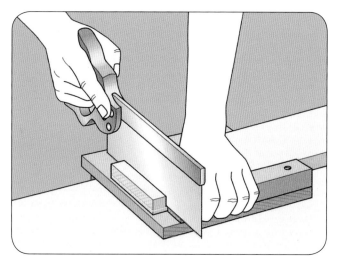

3 First use a tenon saw to cut down to the line, working with the grain and keeping the blade on the waste side of the line. Then, with the piece held fast against a bench hook, use a back saw to crosscut down to the lap line and remove the waste.

4 Clean up both sawn edges with a shoulder plane, taking care to keep the cuts even along their full length. Test the fit of the joint at intervals, continuing to adjust with a shoulder plane if necessary.

Cutting mitred lap joints

This follows the same steps as for cutting a standard lap joint, but the end of the lap comes to a point.

1 Follow the steps for marking a standard lap joint, but mark a 45-degree line at the end of the lap. Continue the line on the inside face of the lap with a square, and mark a mitre on the other side so you have three pencil lines to work to. Support the work in a vice with waste material beneath the lap to avoid tear-out. Use a piece of wood with its end bevelled to 45 degrees as a guide for the plane to keep the mitre true. Use a block plane to create the mitre, but be careful not to cut beyond marked lines.

2 Take the second workpiece and use a mitre square to mark a 45-degree line from both external corners to form the second mitre. Now use the first marking gauge – set to the thickness of the lap – to mark the three shoulder faces, starting at the first mitre line and finishing at the second.

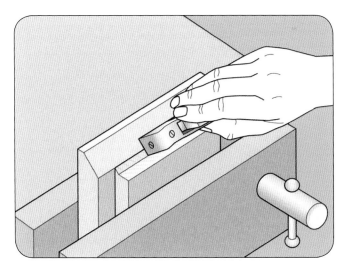

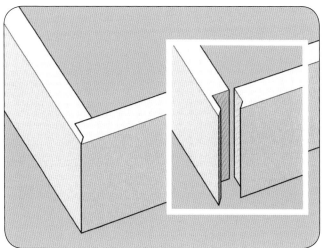

3 Place the workpiece in a vice and cut close to the mitre line with a back saw to remove the waste. Next hold the workpiece firmly against a bench hook and crosscut close to the line until you reach the first cut – this last stage requires a good deal of care. With a shoulder plane remove any waste left by the saw, trimming down to the scribe line.

4 The mitre is also trimmed with a shoulder plane, although a wide chisel may be needed to clean up the section where the mitre meets the flat of the shoulder.

Avoiding tear-out

If you don't have a section of waste material the right size to support your work to avoid tear-out, use a block plane from both sides of the lap in turn, stopping just past the midpoint. Extra care will need to be taken to keep the mitre accurate, but once mastered this is a great technique to have available.

Scarf joints

In most circumstances it is more desirable to obtain wood of sufficient length than to join pieces end to end, but the scarf joint is always available should the need arise. It consists of a taper that is cut and then planed flat on both pieces; the taper length should be at least four times that of the thickness of the wood to give the joint a large gluing area, which increases the strength of the joint. Once glue is applied cramping (clamping) can be difficult because there is a tendency for the two tapers to slide away from each other; to avoid this, cramp (clamp) the two pieces to a flat surface to keep them in position. Avoid glue squeezing out by applying just enough glue to lightly cover the joint area.

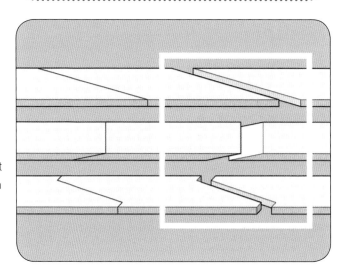

housing (dado) joints

Housing (dado) joints are very useful in shelf and cabinet construction. In its most basic form the through housing (dado) joint provides good shelf support when used within an already rigid structure. However, without a rigid structure surrounding it a standard housing (dado) joint will not provide much strength under tension because – given its small long grain gluing area – the joint can be pulled apart relatively easily. A stopped housing (dado) – where the joint is only partially cut through the width of a board – disguises one side of the joint, making this an attractive option for exposed areas of furniture. Housing (dado) joints can also be cut with shoulders, which requires extra work but can be visually beneficial because the shoulders disguise any discrepancy in fit over the length of the housing (dado).

Cutting housing (dado) joints by hand

These instructions are for a housing (dado) joint for a shelf.

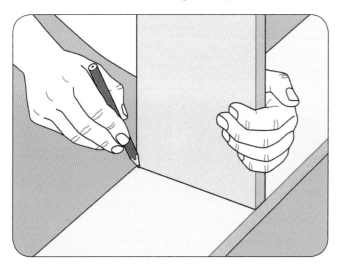
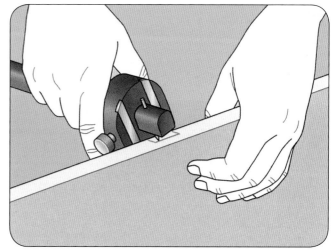

1 Start by crosscutting the male part of the housing (dado) joint square – in this case the end of the shelf. Next mark the uppermost line of the housing (dado) across the piece that will feature the housing (dado) cut – in this example, mark the position of the top surface of the shelf.

2 Take the male component (the shelf) and marry it up with the line, and then mark the thickness of the shelf by scribing a line along the bottom edge. Remove the shelf and mark the depth of the housing (dado) cut, ensuring it does not exceed two thirds of the material thickness.

> **Related info**
>
> Choosing the right joint (see page 180)
> Marking out (see page 181)
> Wood characteristics (see page 124)
> Adhesives and assembly (see pages 214–223)

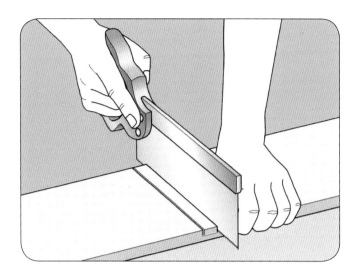

3 Use a tenon saw to cut both sides of the housing (dado), cutting with long strokes. For extra support cramp (clamp) a batten in place to guide the saw.

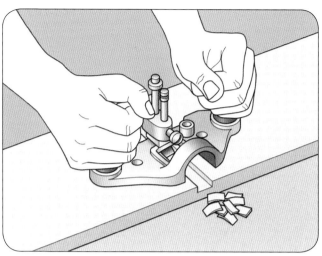

4 Remove the waste material with a router plane. A paring chisel also works well, but requires more skill to accurately gauge the depth of the cut.

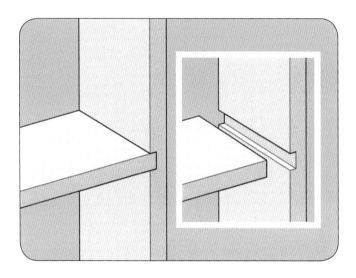

5 Try the male part of the joint for fit. If necessary, trim the housing (dado) a little more with a chisel until you achieve the desired fit.

Cutting housing (dado) joints by machine

A router fitted with a straight flute cutter is used to cut housing (dado) joints. Mark out the housing (dado) as in steps 1–2 of cutting by hand. If the male part has shoulders, the thickness of the tongue can be matched to the diameter of the cutter. If not, line the cutter up on both sides of the housing (dado) in turn to achieve the correct width. Set the depth of cut to match the housing (dado) depth. Use a straightedge as a guide for the router (see box Positioning a straightedge to guide the router on page 192). With the straightedge cramped (clamped) parallel to the cut line, remove waste material from the housing (dado) in a series of light passes. If the housing (dado) is wider than the cutter, repeat this step for the second edge.

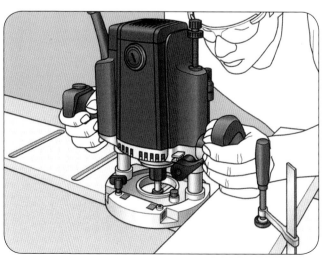

Positioning a straightedge to guide the router

When using a straightedge to guide your router it is crucial to know the measurement from the centre of the router cutter to the edge of the router base plate.

1. Take a piece of material with a square edge and mark a pencil line across it with a try square.
2. On the router base there should be a vertical indentation marking the centre at both front and back of the base. Line these marks up with your pencil line.
3. Use a pencil to mark around the router base on either the left or right side. The edge of the router base may be straight or curved depending on the model.
4. Remove the router and measure the distance at a right angle from the centreline to the furthest point of the base. This measurement should be the distance between the centre of the cutter and the edge of the base – it will be different on each router so for the next step, this measurement is referred to as X.
5. Divide the cutter diameter in half to establish the cutter radius. To work out how far the straight-edge should be from the edge of the cut line, use the following formula: X minus cutter radius equals distance between straightedge and cut line.

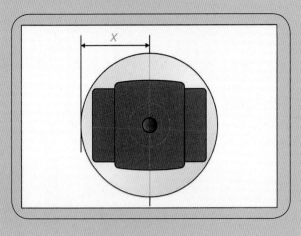

Cutting housing (dado) joints

A sliding compound mitre saw (SCMS) with trenching facility or a radial arm saw will both cut housing (dado) joints very well. These machines can also be set to cut shoulders accurately.

For situations that need added strength consider using a dovetail housing (dado) joint.

Dovetail housing (dado) joints by hand

The dovetail housing (dado) is a difficult joint to cut by hand, but with care and practice it can be done very well.

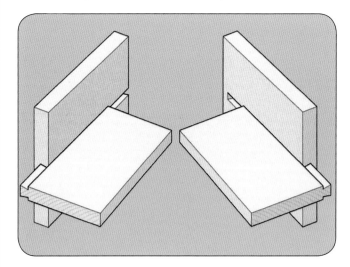

1 Establish an appropriate size of dovetail housing (dado); the depth of the housing (dado) should not exceed two thirds of the wood thickness and the width of the narrowest part of the housing (dado) should not be less than one third of the thickness of the board that features the male part of the joint.

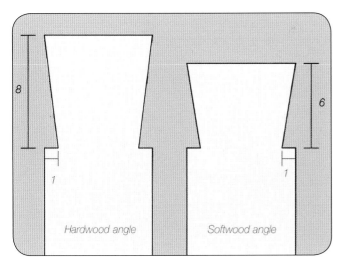

Hardwood angle   Softwood angle

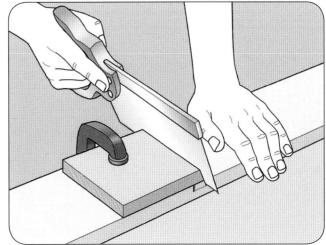

2 Establish the position of the housing (dado) by marking a centreline across the width of the board with a try square. Mark two additional lines at equal distances from the centreline to form the width of the dovetail housing (dado) at its narrowest point. Take a sliding bevel and set this to 1:8 for hardwood or 1:6 for softwood (see box Setting a sliding bevel for dovetailing on page 194), then mark the sides of the dovetail housing (dado) onto the edge of the board, stopping at the depth line. If cutting a through dovetail housing (dado), make sure both ends of the housing (dado) are marked. Before cutting, crosshatch the waste area clearly with a pencil.

3 Take a straight wooden batten and cramp (clamp) it along the cutting line to the left of the centreline to form a fence to guide your tenon saw. Take long gentle cuts the full width of the housing (dado), while keeping a close eye on the angle of the saw blade. Sight down the cut line at regular intervals to ensure you are cutting in line with the angle of the dovetail. Keep sawing until you reach the full depth of cut at both ends of the housing (dado), then repeat this process for the other side.

Achieving a fine cut

When using a paring chisel to remove waste material from the housing (dado), use your left hand (right hand if you are left-handed) to guide the blade for greater control. This will enable you to achieve a much more accurate and finer cut.

4 Use a router plane to remove the waste in stages. Rotate the router plane sideways slightly to get into the corners of the dovetail housing (dado). Depending on the length of your housing (dado) you may find that a long narrow paring chisel is ideal to remove the last areas of waste at this stage.

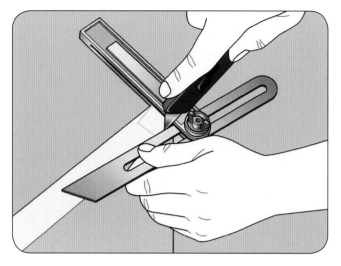

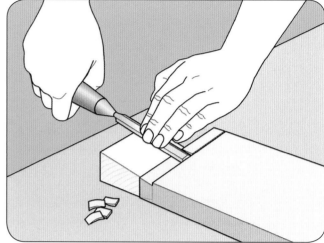

5 To cut the male part of the joint, mark it out with the sliding bevel and try square to fit snugly into the housing (dado). Use a marking knife to mark in the shoulders and then crosscut these with a back saw.

6 Use a combination of shoulder plane and paring chisel to gently remove the waste. Test the joint for fit and keep adjusting as necessary by removing thin shavings of material, until it fits well along its entire length.

Setting a sliding bevel for dovetailing

To set a sliding bevel for marking dovetails in hardwood, take a piece of material with a square edge and draw a line at right angles to the edge, then mark a point on the line 200mm (8in) from the edge. Measure 25mm (1in) along the edge from the base of the line, and then draw a third diagonal line that joins the two points to form a scalene triangle. Place the sliding bevel on the edge and set it to the diagonal line – this is the correct angle for cutting dovetails in hardwood and is referred to as a 1:8 angle. For softwood repeat the process above but using 150mm (6in) instead of 200mm (8in) to create a 1:6 angle.

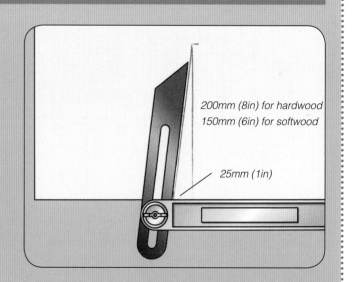

200mm (8in) for hardwood
150mm (6in) for softwood

25mm (1in)

Dovetail housing (dado) by machine

Since a dovetail housing (dado) is not easy to cut by hand, it is more common to cut them using a router. Choose an appropriate size dovetail cutter for your router. Mark out the dovetail housing (dado) by first drawing a centreline where the centre of the housing (dado) is to be. Draw parallel lines either side of the centreline, spaced so that the finished dovetail housing (dado) will be slightly wider than the cutter – this allows you to concentrate on cutting one side of the housing (dado) at a time to reduce stress on the cutter. Finally mark the depth of the housing (dado), to be not more than two thirds of the depth of the material. Fit a straight flute cutter to the router, using a diameter smaller than the narrowest width of the dovetail. Line up and secure a straightedge to guide the router parallel to the housing (dado), so the flute cutter can remove the majority of waste without passing over the marked lines. Set the depth of the router cutter just above the depth mark of the dovetail housing (dado) – for a deep housing (dado) you may need to make two passes.

Carefully run the router along the straightedge to remove as much waste as possible. Having removed the majority of waste, fit a dovetail cutter in the router and realign the straightedge (see box Positioning a straightedge to guide the router on page 192) for one side of the dovetail housing (dado). Check the alignment

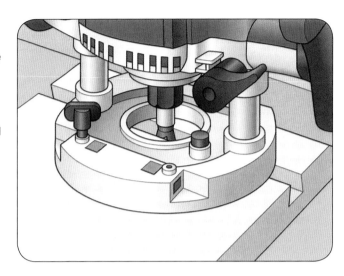

and cutter depth then carefully make the cut. Repeat the process for the other side of the dovetail housing (dado). To make the male part of the joint, use a router table. First test the settings by cutting the dovetail profile on both sides of a test piece exactly the same thickness as the final material, and then try this for size. Repeat this process until a good fit is achieved – ensure the fit is not too tight because this will prove very hard to push home, especially on a wide dovetail housing (dado).

Fitting a dovetail

Making a perfect fitting dovetail housing (dado) can be quite difficult. A slightly looser fit is often best, because friction will increase as the joint is slid together.

biscuit, dowel and pocket screw joints

Biscuit joints work well in both solid wood and manufactured boards. They offer a quick and highly efficient method of jointing – especially when a dedicated hand-held biscuit jointing machine is used to cut the slot.

Biscuit jointing

Biscuits are made of compressed wood and are available in a range of pre-cut sizes. Because the biscuits are compressed they expand when glue is applied, which creates a very tight joint.

Related info

Choosing the right joint (see page 180)
Wood characteristics (see page 124)
Biscuit jointers (see page 77)
Marking out (see page 181)
Adhesives and assembly (see pages 214–223)

Cutting an edge-to-edge biscuit joint

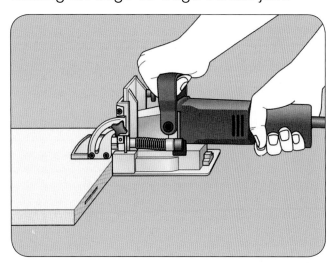

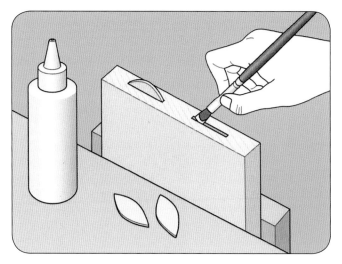

1 Align the two pieces to be jointed and then use a pencil to make a small mark at right angles across the join line. Repeat for the placement of each biscuit. Set the biscuit jointer to the correct depth setting to match your choice of biscuit size. With the workpiece firmly held in place, line up the centre of the biscuit jointer with each mark and then plunge the machine into the workpiece to form the biscuit slot.

2 Glue and cramp (clamp) the joint – remember that only a small amount of glue is needed in each slot because the fit will be tight and excess glue will squeeze out.

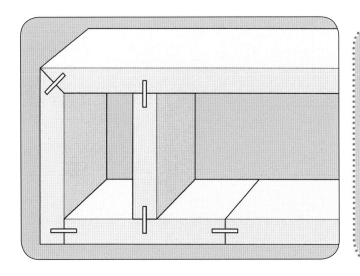

Cutting slots

If you do not have a dedicated biscuit jointer you can use a slot cutter installed in a router table. While still efficient this is a slower method of biscuit jointing because additional time is required to mark the correct length of each slot, unless a dedicated stop system is used.

3 As well as edge-to-edge joints, right-angle and mitre joints can also be created with a biscuit jointer. This through-section shows some instances where a biscuit joint can be used.

Dowel jointing

Because of its length, a dowel joint provides a similar level of constructional strength to a mortise and tenon joint. Dowel joints are easy to use and well suited to a variety of tasks, including cabinet and chair construction. Because of all these factors this joint is a good choice for both the novice and experienced woodworker. In its most basic form a set of dowel points and bits is used to accurately drill and place dowel holes relative to each other.

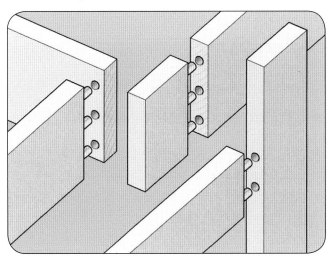

Making dowels

If you make your own dowels make sure that you cut a groove along the length of each with a tenon saw to relieve the hydraulic pressure that builds up as the dowel is hammered home.

construction methods

Making a dowel joint

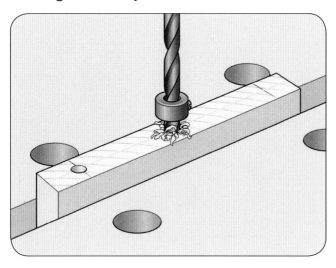

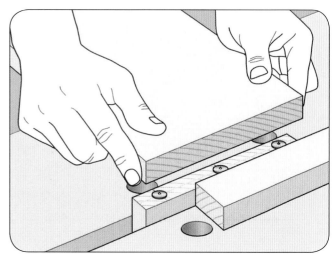

1 Mark the centre position of each dowel hole on one component. Hold the work in a vice, then drill a hole of the correct diameter and depth for each dowel. Sight down the bit as you drill, keeping as square to the face as possible. If the drill does not have an adequate depth stop fitted, a piece of masking tape wrapped around the drill bit will help you work to the correct depth.

2 Insert a dowel point into each hole, then bring the two components together and apply light pressure so that the points leave a slight indentation. Drill the second set of holes, then glue lightly before cramping (clamping) the joint evenly and setting aside to dry.

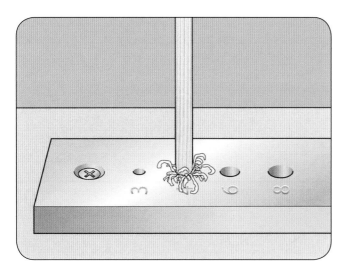

3 Dowels can be bought cut to size or you can make your own using a dowel plate, which is a heavy gauge metal plate with pre-cut holes for forcing wood through and profiling it to size. Various jigs can also be used to achieve accurate positioning of dowels.

Pocket hole screw joints

Pocket hole screwing is a versatile form of joining both wood and manufactured boards together; joints are strong, convenient and easy to create using a dedicated jig. Like biscuit and dowel jointing, pocket hole joints are ideal for beginners and quick projects. Pocket holes are drilled from one side of the joint, so careful planning is needed to hide them from view. The holes can be plugged with lengths of dowel, but the jointed side will always be noticeable to the trained eye unless the surfaces are painted.

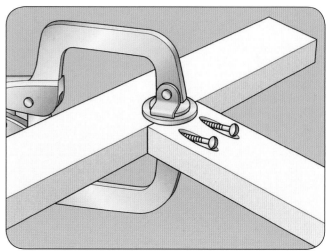

Miller dowel system

The Miller dowel system provides an easy and efficient form of jointing components together. A special stepped drill bit is used to bore the holes, which correspond to one of several sizes of Miller dowel available.

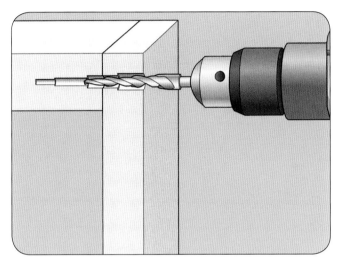

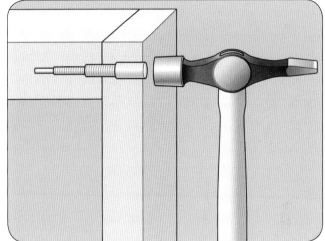

1 Temporarily hold the two components to be joined together with cramps (clamps), so that both your hands are free. Use a dedicated Miller dowel bit in the drill to bore a stepped hole through both components simultaneously.

2 With the components still cramped (clamped) in position, apply a little glue to a Miller dowel and tap it into place with a small hammer. Set the joint aside to dry. When the glue has set, saw off any remaining dowel and then trim the end flush with a block plane or chisel.

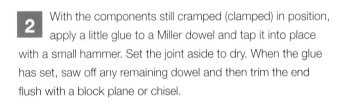

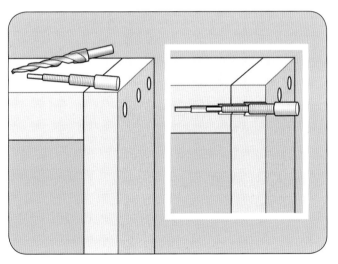

3 Miller dowels are available in a variety of woods so they can either be matched to the main wood or used in a contrasting colour for visual effect.

mortise and tenon joints

The mortise and tenon joint is traditionally used in frame and panel construction to create large panelled areas; the frame holds the solid wood panels isolated so they are free to expand and contract with the seasons. Mortise and tenon joints are widely used in other forms of construction too, such as table/desk under-frames and for chairs.

Making mortise and tenon joints

There are many ways to make mortise and tenon joints, but regardless of the method always start by cutting the mortise because it is easier to trim the tenon to fit the mortise rather than the other way around. For speed of production, the mortise is usually cut to the width of a chisel, although this should not exceed one third of the width of the wood. Height is determined by the height of the rail to be tenoned, and mortise depth is generally between half and two thirds of the thickness of the wood to be mortised. If you regularly cut mortises in wood of similar width and thickness, then consider buying a mortise chisel or two to match the dimensions required.

When cutting the mortise, some woodworkers prefer to chop directly to the line at each end first, ensuring that ends are dead square (see page 111). This is because if there is too much slack in the sliding table, plunging the chisel down with a few millimeters left can force the piece to move slightly sideways, resulting in out-of-square ends.

Haunched mortise and tenon joints are used where a rail is flush with the top or bottom of the stile. The shallower haunch maintains joint integrity by minimizing end grain tear-out while keeping the tenon as tall as possible to reduce the risk of twisting over time. If two mortise and tenon joints are to be cut for a corner – a good example being a table leg where two rails meet – mitre the ends of each tenon to allow clearance for the adjacent tenon. If the mortise and tenon joint is to form part of a frame and panel construction, a shallow groove is run along the stiles, muntins and rails, in line with the mortises, to accept a panel.

Marking out haunched mortises

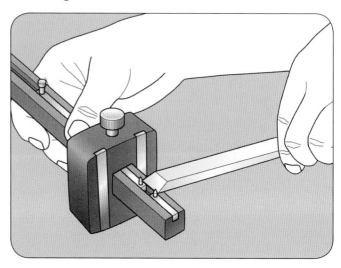

1 Set a mortise gauge to the width of the selected mortise chisel and centre the marks to the width of the workpiece by marking from both sides and then adjusting the gauge so that both sets of marks meet exactly.

Related info

Choosing the right joint (see page 180)
Wood characteristics (see page 124)
Mortise and tenon jigs (see page 87)
Mortisers (see page 110)
Marking out (see page 181)
Adhesives and assembly (see pages 214–223)

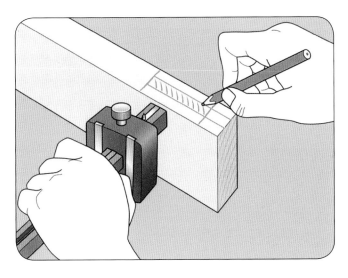

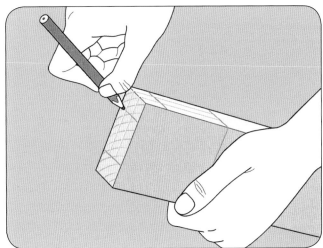

2 Mark the height of the mortise with a pencil and square. Include a squared mark for the start point of the haunch. Use the mortise gauge set in step 1 to mark the sides of the mortise between the squared lines, marking the extremities of the mortise.

3 Make two depth marks on the end grain of the workpiece to be mortised. These will be used as a depth reference when cutting the mortises.

Cutting mortises by hand

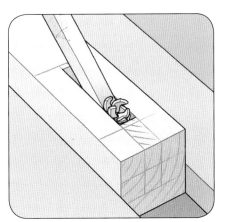

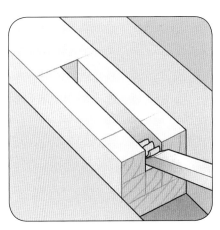

1 Keeping a small distance away from the start and end marks of the mortise, use a mortise chisel and mallet to remove the waste one layer at a time. Work the length of the mortise taking care not to over-cut the depth of the haunched area.

2 When the approximate depth of the mortise has been reached, clear out any waste and trim back to the squared lines that were made in step 2 of marking out haunched mortises.

3 Trim the haunched area back from the end of the wood so that its base is flat and to the correct depth. Check the internal measurements of the mortise and adjust if necessary.

Cutting mortises with a mortising machine

The hollow chisel mortiser offers one of the quickest methods of cutting mortises by machine. When cutting mortises with a router the corners are round, but the hollow chisel mortiser cuts mortises with square corners so no additional trimming of the tenon is necessary.

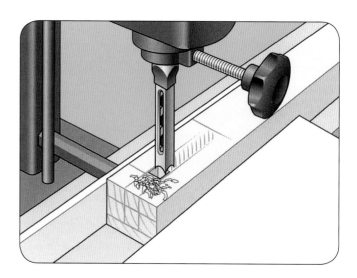

Fit the mortiser with an appropriate size chisel and bit, and then set the depth of cut using the depth marks created in step 3 at the top of page 201. If your machine is fitted with a dual depth stop, the haunch depth can be set at the same time. Set the machine up so that the chisel is lined up with the mortise area, then take the first cut away from either end of the mortise. The first cut should always be made in stages: plunge the chisel in to approximately one quarter depth, then withdraw to allow the chips to clear; plunge again to half depth; repeat this process until full depth is reached. Next, work the length of the mortise by plunging the chisel so only half of its width is chopping out material – this prevents overloading. Repeat this process until all waste is removed from the main mortise area. Reset the depth of cut and remove waste from the haunch area.

Cutting tenons by hand

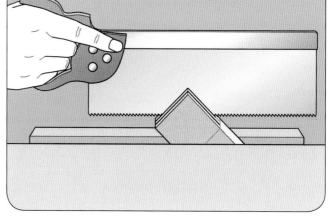

1 Using the same mortise gauge settings, mark the tenon on all three sides. Pencil in the shoulder lines on all four faces and then use a knife and square to mark in the shoulder lines. Place the workpiece in a vice at approximately 45 degrees then proceed to cut from the marked corner with a tenon saw. Keep the saw on the waste side of the line – better the tenon is too tight at this stage than too loose. Repeat this process for the adjacent tenon shoulder.

2 Rotate the workpiece and repeat the process to cut away on the opposite side.

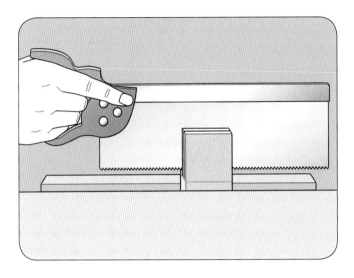

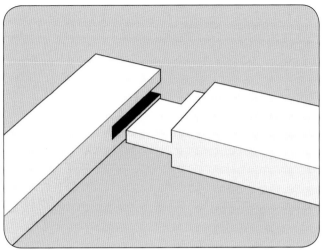

3 Place the wood square in the vice and cut out the triangle left in the middle. Use a bench hook to hold the wood in place while you crosscut the shoulders away. With the workpiece upright in the vice and relatively close to the bench top, use a wide bevel edge chisel to pare the shoulders back to the marked lines. Mark the haunched area with a small square and steel rule and then cut away the waste with a tenon saw.

4 Try the tenon for fit. If it is too tight use a shoulder plane or wide chisel to gently pare material away in equal amounts from both sides of the tenon, then retry the fit. Repeat this process as necessary until a good snug fit is achieved.

Cutting tenons by machine

There are many ways to cut tenons by machine but this tried and tested method is quick and provides reliable results that can be adjusted and repeated as needed. It is best to perform this sequence on a piece of scrap material first to test the fit.

A bandsaw is used to cut the tenon cheeks; use the fence to line up the blade with the marked line closest to the fence. Cut the first cheek and then rotate the workpiece and cut the second cheek. Use a mitre saw equipped with a trenching facility, or a radial arm saw with its depth of cut set, to cut the shoulders. Test the fit and, if necessary, tweak the position of the bandsaw fence so that you can cut perfect tenons time and time again. It is worth removing the haunch material by hand with a tenon saw, because this allows the bandsaw to remain set for cutting additional tenons.

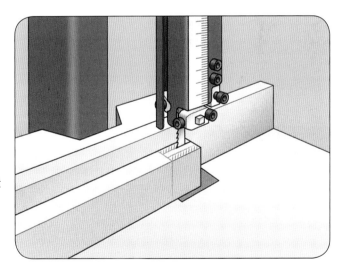

Mortise and tenon options

There are many variations on the mortise and tenon joint and many projects that you undertake will require a slightly different joint each time. Don't be afraid to modify the standard mortise and tenon joint to your requirements.

Through mortise and tenon joints Mortises can be cut all the way through a section of timber (lumber), although it is better to cut just over halfway and then complete the cut from the other side to prevent tear-out. The tenons – cut slightly over long – are then glued in place and trimmed back flush with a block plane.

Wedging joints Add wedges to tenon ends to create exceptionally strong joints. To add wedges, cut a slight bevel at either end of the mortise on its outside face to allow the tenon to expand. Use a tenon saw to cut two lines vertically down the length of the tenon. Make up two appropriate-size wedges from material cut away from the tenon shoulders. After the joint is assembled and cramped (clamped) in place, drive the wedges home. This effectively expands the joint to form a dovetail profile and increases its strength considerably.

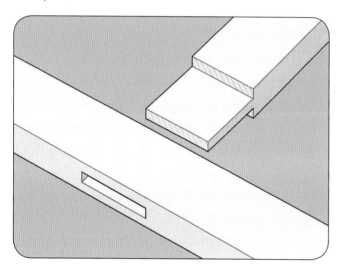

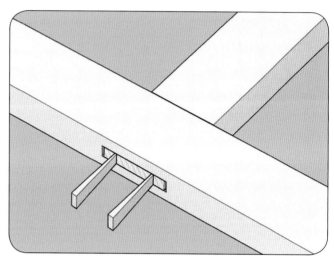

Offset joints Sometimes it is necessary to create mortise and tenon joints off-centre – a good example of this would be a rebated (rabbeted) frame.

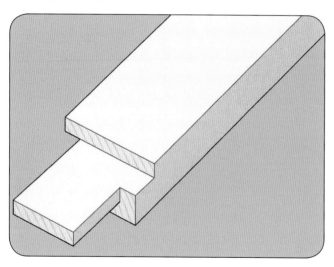

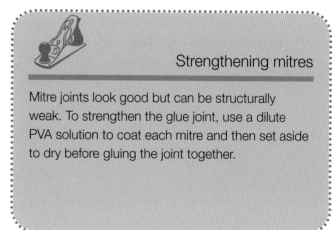

Strengthening mitres

Mitre joints look good but can be structurally weak. To strengthen the glue joint, use a dilute PVA solution to coat each mitre and then set aside to dry before gluing the joint together.

halving and bridle joints

Halving joints provide a straightforward but effective method of jointing wood, and are useful in a broad range of situations. Often referred to as slot or open mortise and tenon joints, bridle joints are easy to produce and used extensively in framing.

Halving joints

In this joint, two components – each with half their thickness removed at the joint – are brought together, and then glued and cramped (clamped) to create a joint that is a single thickness deep.

Related info

Choosing the right joint (see page 180)
Marking out (see page 181)
Wood characteristics (see page 124)
Adhesives and assembly (see pages 214–223)

Cross halving joint

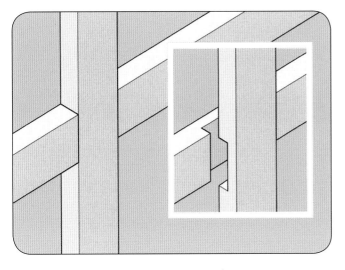

T-halving joint

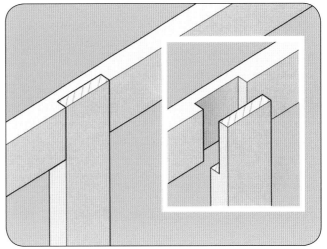

Mitred corner halving joint

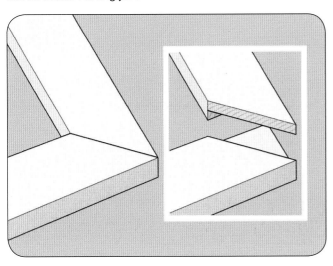

Corner halving joint

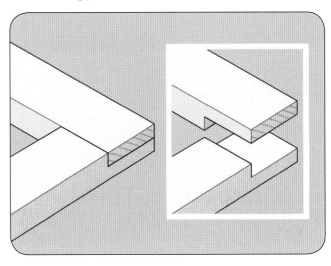

Glazing bar halving joint

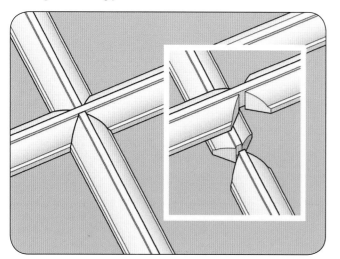

Dovetail halving joint

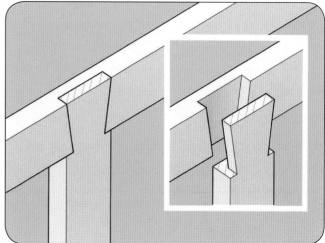

Cutting halving joints

Halving joints can be cut with a mitre saw equipped with a trenching facility or a radial arm saw set to the correct depth of cut. Successive cuts are taken across the width of the workpiece until all the waste material is removed. Alternatively a router can be used against a guide rail to achieve the same result.

Making a cross halving joint

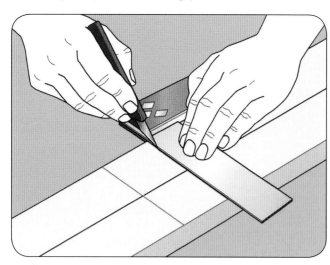

1 Use a square to mark where the halving joint will begin on the first workpiece. Use the second workpiece as a guide to mark its exact width onto the first workpiece. Take a marking knife and score the two crosscut lines.

2 Set a marking gauge to half the thickness of the wood. Mark the depth of the cut with the marking gauge.

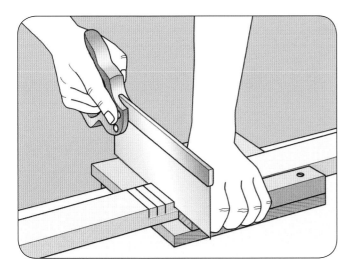

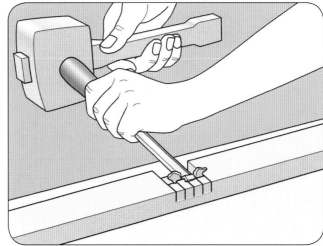

3 Place the workpiece on a bench hook and then – starting inside the first crosscut line – make a series of cuts across the workpiece down to the gauge line, finishing just inside the second crosscut line.

4 With a wide bevel-edge chisel, chop out the waste from both sides and then pare back to the gauge line. Pare the shoulders back flush with the knife lines using the chisel. Repeat this sequence on the second workpiece, then bring the two components together to check the fit.

Bridle joints

The bridle joint is usually divided up into three parts of equal thickness – two cheeks and one tenon.

Making a corner bridle joint

In these steps it is assumed that both workpieces are of equal thickness.

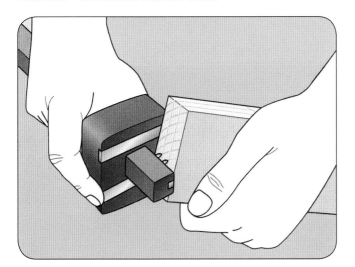

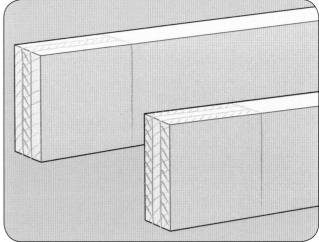

1 Set a mortise gauge to mark the central section of the workpiece. Check that the gauge is set up accurately by comparing it from opposite sides of the workpiece to see if the points line up correctly.

2 Mark both joint components with the gauge, then use a square and marking knife to mark the shoulders of the tenon and the depth of the open mortise. Use pencil for lines that may be visible on outer surfaces and to crosshatch waste areas.

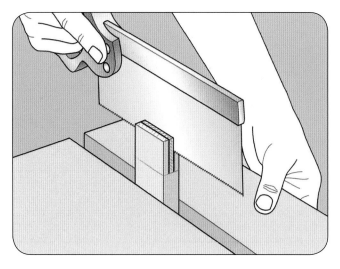

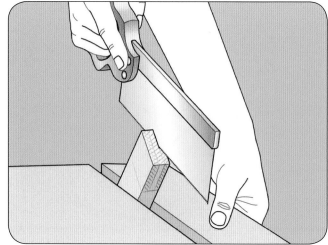

3 Cut the open mortise section of the joint first. Place the workpiece in a vice at approximately 45 degrees and then cut from the marked corner with a tenon saw. Keep the saw on the waste side of the line. Repeat this process for the adjacent gauge line. Rotate the wood and repeat the process on the other side. Next, place the workpiece vertically in the vice and cut out the triangle that is left in the middle by sawing at right angles to the workpiece. Remove the central waste section with a coping saw and then pare back to the line with a bevel edge chisel.

4 To cut the tenon part of the joint, repeat step 3 but remember that the waste is reversed – instead of removing the central section to form the cheeks, the central section will be the tenon so you must remove the cheek areas to form the shoulders. Cut the shoulders using a bench hook and back saw, and then pare back to the line with a wide bevel edge chisel.

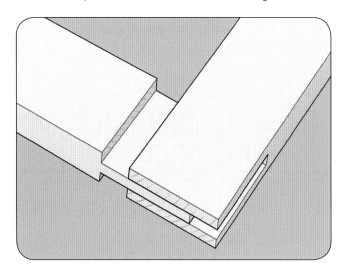

5 Check the fit – if the joint is too tight gently pare material away from the tenon in equal amounts on both sides to ease as required.

Cutting bridle joints

If both workpieces are not of equal thickness, then the marking gauge must be used from one set of faces only – usually the outside faces where both components are flush.

To cut the joint by machine, for rip cuts where a tenon saw is used in the steps, a bandsaw can be substituted with the fence set to cut the correct thickness of cheek and shoulder. To remove shoulders, use a mitre saw with a trenching facility or a radial arm saw.

dovetail joints

The dovetail joint is traditionally used in drawer construction because the wedge shape tails make it particularly strong. Dovetails are also used extensively in carcass construction and box making projects. Since numerous dovetail jigs are available commercially for making dovetail joints with a router, the following steps are all shown using hand tools. For softwoods dovetails should be cut at an angle ratio of 1:6, while for hardwoods they are cut at 1:8 – see box Setting a sliding bevel for dovetailing on page 194.

Through dovetails

For traditional drawer making and many types of chest and box projects, through dovetails offer an elegant solution with plenty of structural integrity.

Making through dovetails

Whether you cut the tails or pins first is down to preference – in Britain the tails tend to be cut first whereas in the United States the pins are usually cut before the tails. The steps here are all based on cutting the tails before the pins.

Related info

Choosing the right joint (see page 180)
Wood characteristics (see page 124)
Setting a sliding bevel for dovetailing
(see page 194)
Dovetail jigs (see page 87)
Marking out (see page 181)
Adhesives and assembly (see pages 214–223)

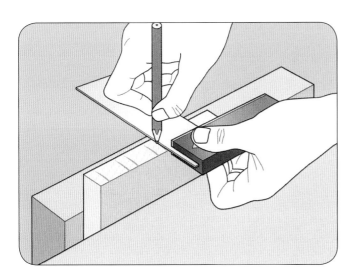

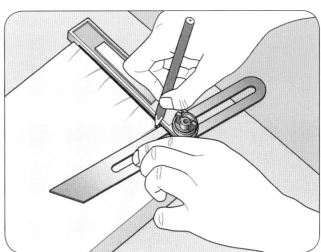

1 Work out the spacing of the dovetails – 6mm (¼in) clearance at either end is generally a good distance. Mark the position of the dovetails across the end grain of the tail board (the workpiece that will have the tails cut in it) with a pencil and square, leaving a 3–6mm (⅛–¼in) gap between each tail.

2 Use a sliding bevel set to either 1:8 or 1:6 to mark in the tails on both sides of the tail board. There is no need to knife in these lines – pencil marks are fine at this stage.

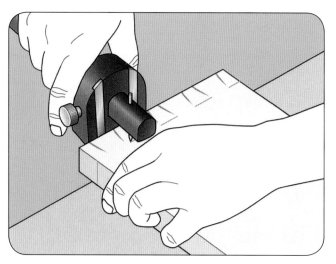

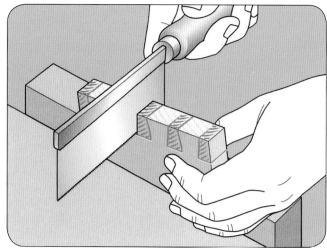

3 Set a marking gauge to the thickness of the pin board (the workpiece that will have the pins cut in it) then gauge a line across all the waste areas between each of the tail marks. Repeat this step on the opposite side. Crosshatch the waste areas to reduce mistakes when it comes to the cutting stage.

4 Place the tail board in a vice with its end fairly close to the bench top to minimize flexing. It can either be secured vertically so that the tail lines are at an angle, or at an angle so that half the tail lines are vertical – whichever you are more comfortable with. If the latter method is chosen the tall board must be repositioned to cut the second lot of vertical cuts. With a dovetail saw cut the waste side of each line. Start the cut with the saw angled up slightly so that you are cutting a comer then – once the saw has started to cut easily – straighten it out so that the marked tail line can be accurately followed. Repeat this step for all of the angled tail lines, but make sure that you do not cut past the depth line created with the gauge in step 3.

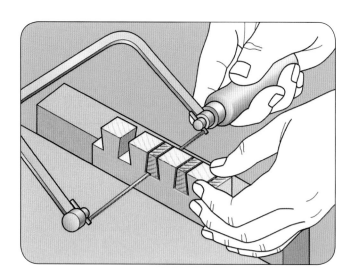

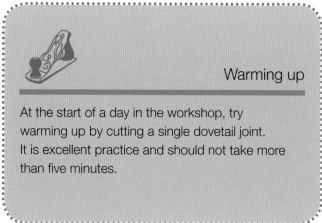

Warming up

At the start of a day in the workshop, try warming up by cutting a single dovetail joint. It is excellent practice and should not take more than five minutes.

5 Remove the waste with a coping saw. For this step the tail board must be secured vertically in the vice. Again, keep the cutting area close to the bench top to minimize flexing.

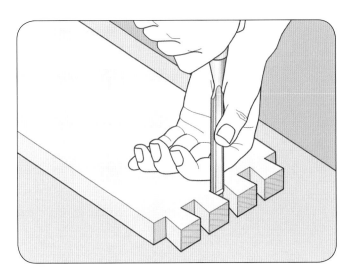

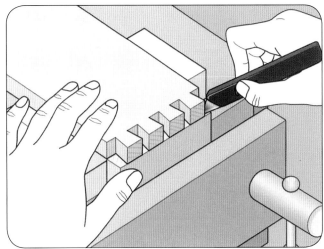

6 With the bulk of the waste removed, lay the tail board flat on a piece of scrap material on the bench and cramp (clamp) securely in position. Use a bevel edge chisel to pare the remaining waste back to the gauge line between each set of tails. Work in stages, removing small slivers at a time before cutting as exactly to the gauge line as possible. When paring away the waste cut to an approximate halfway point – never cut all the way through the full thickness of either the tail board or pin board because this risks tear-out on the opposite side. Rotate the work and complete the paring cuts from the opposite side.

7 The next step is to mark the pins directly off the cut tails. Position a piece of scrap material on the bench top just behind the vice, and place the tail board on top of it so that the tails protrude over the vice opening. Place the pin board with its pin end directly below the tails and secure in the vice so that both workpieces are square to one another. Now use a marking knife to mark the outline of the tails onto the end grain of the pin board.

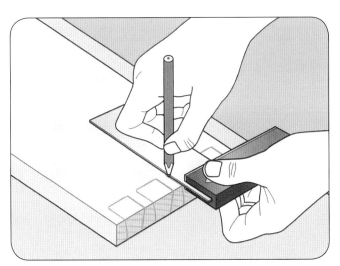

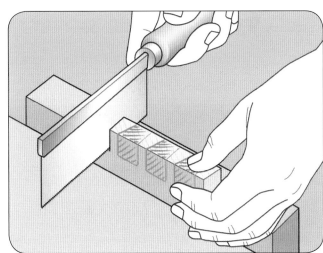

8 Remove the pin board from the vice and mark in the waste areas with the marking gauge and square – using the settings that have already been established in step 3 – then crosshatch the waste so you know which side of the line to cut.

9 Secure the pin board vertically in the vice, then proceed to cut on the waste side of each line, keeping your cuts as close to the line as possible – the aim is to fit the dovetails directly off the saw.

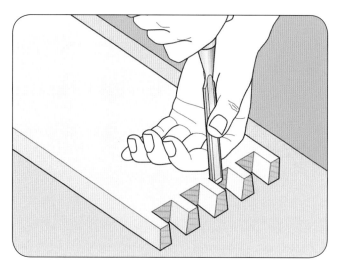

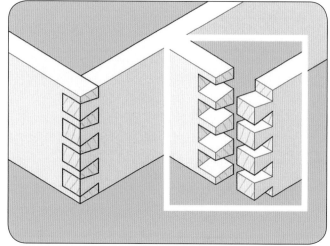

10 When all the vertical cuts are complete remove the waste with a coping saw. Using the same procedure outlined in step 6 on page 211 to pare down to the gauge line between each set of pins.

11 Try the dovetails for fit. Some trimming may be necessary with a chisel to achieve a good fit but with practice you will be able to fit straight from the saw. When test fitting dovetails never push them fully home – only go halfway to avoid weakening the joint.

Fitting from the saw

Try to make your dovetail joints fit just right straight from the saw. This approach will save time and will also improve your cutting skills for other tasks.

Lapped dovetails

Lapped dovetails are primarily used for drawer fronts where it is not desirable to see the ends of through dovetails. They are also used in carcass construction, where cross rails need to be jointed to the tops of desk legs or corner stiles in frame and panel construction.

Making lapped dovetails

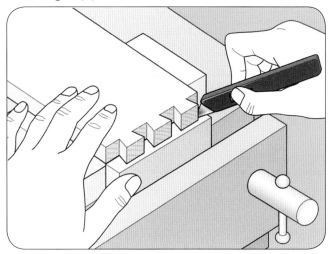

1 Mark out and cut the tails in the same way as outlined in steps 1–10 of making through dovetails on pages 209–212 but instead of setting the marking gauge to the full thickness of the pin board, set it to three quarters of the thickness. When marking the pins from the tails mark the ends of the tails as well, because they will be set in by one quarter of the pin board thickness.

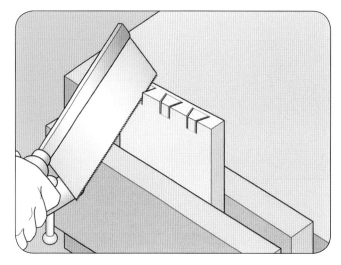

2 Reset the marking gauge to the full thickness of the tail board then gauge a line across the inside face of the pin board. Use a square to mark in the vertical lines between the tail marks and the gauge line. Secure the pin board vertically in the vice then proceed to make angled cuts with a dovetail saw to the extremities of both tail marks and gauge line.

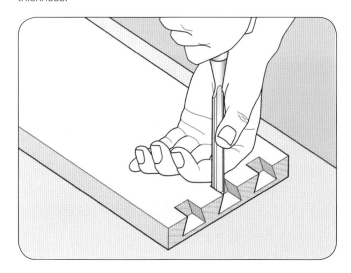

3 Cramp (clamp) the pin board on a piece of waste material on the bench top. Remove most of the waste by striking the chisel lightly with a mallet to make a series of stepped cuts. Finally pare into comers and to the marks with a sharp bevel edge chisel.

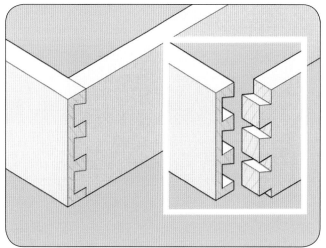

4 Test the fit of the joint and adjust with the chisel as necessary if it is too tight.

adhesives and assembly

Adhesives have come a long way in recent years; modern formulations provide exceptional strength and versatility while creating bonds that are often stronger than the wood being joined. This chapter will guide you through the process of choosing the correct adhesive and assembling your projects for best results.

preparation

Gluing-up is often one of the most stressful stages of a project. All that hard work is at stake; an error at this stage could be awkward to put right and on top of that the clock starts ticking the minute you start applying glue.

Minimizing glue-up problems

Before joints can be glued together surfaces need to be clean and free of dust or substances that might compromise the adhesive's ability to form a strong bond with the wood. Freshly cut wood is an ideal surface, but if you are regluing a joint then old adhesive should be removed prior to applying fresh glue.

Before gluing-up try fitting all the components together dry individual joints will often join adequately but when a project is assembled as a whole, with several joints being brought together simultaneously, problems can arise. These might not otherwise have become obvious until gluing-up for real – by which time it is often too late to put them right – but by going through a dry run most problems can be remedied prior to applying any glue. Most joints can be dry assembled, although dovetail joints should only ever be partly assembled – especially if a good fit has been achieved – to avoid any unnecessary loosening of the joint prior to gluing.

At this stage decide which parts of your project should have finish applied before and which after gluing-up. Best results are often only possible by applying finish to interior faces before gluing-up because access – especially to corners – can be compromised once the item is assembled.

Choose your glue carefully and make sure that you have enough open time – the amount of time you have to work with the glue until it sets – for your project. Many types of glue are available in alternative formulations with different lengths of open time, so if you favour a particular type of glue for general woodwork, consider keeping stock variants with both long and short open times. Long open time is ideal for complex assemblies, while short open times are convenient for the small gluing jobs required during a project build – for example, gluing narrow boards together to make wide panels.

Temperature and humidity will have an effect when gluing-up so be sure to check the required conditions for your particular glue of choice; slightly above or below room temperature with good ventilation is generally about right. However, bear in mind that both these factors can affect an adhesive's open time so if you want to maximize the time you have to assemble a project you could reduce the temperature of your workshop by opening windows or doors, or wait until

> ### Related info
>
> Surface preparation (see page 242)
> Types of adhesive (see page 220)
> Cramping (clamping) (see page 218)
> Safer woodwork (see page 22)

a cool part of the day before starting the glue-up. If you use a dehumidifier to control the workshop humidity level, then switching this off in advance will give you a few extra minutes of open time when using water-based glues.

Some adhesives have nasty effects when they come into contact with skin so be sure to check the safety instructions before using any variety of glue and wear the appropriate eye and hand protection. PVA glues (white or yellow varieties) are among the safest and easiest to use, so these should be high on your list of adhesives for general woodwork. Spatulas, brushes and material off-cuts can be used to spread glue over large surfaces and into hard-to-get-at places.

Before finally gluing-up take a few minutes to think everything through. Have you got everything you need to hand? Have you tried a dry run and ironed out potential problems? Have you got enough glue to finish the job? Have you chosen a glue with the optimum open time? This may all seem like common sense, but it is easy to overlook something crucial if you don't take the time to think it through.

As soon as your cramps (clamps) are secured in place, check your project for square. Instead of using a square to do this, use the much more accurate method of measuring diagonals. This can be done carefully with a tape measure, but for best results use a set of measuring sticks.

cramping (clamping)

Cramps (clamps) are essential for most gluing jobs. The old adage "you can never have too many cramps (clamps)" really is true! A good range of G (C), F and sash cramps (clamps) is invaluable.

Use off-cuts of manufactured board – thin hardboard is best – to insert between cramp (clamp) heads and your work to protect it while applying pressure. Double-sided adhesive tape can be used to semi-permanently fix pieces of hardboard to cramp (clamp) heads to make this job easier and speed up assembly times. Where possible alternate the cramping (clamping) pressure – for example, when gluing boards long grain to long grain, sash cramps (clamps) should be placed alternately above and below the boards to equalize pressure and ensure they remain as flat as possible until the glue sets.

Related info

Cramps (clamps)
(see page 36)
Types of adhesive
(see page 220)
Safer woodwork
(see page 22)

Cleaning up

The best method of cleaning up is to avoid making a mess in the first place. Most squeeze out – the process of excess glue being squeezed from a joint – is avoidable by using just enough glue for the joint. Modern woodworking adhesives are exceptionally strong so not very much glue is needed to create a good bond; as long as your joints fit well a thin layer of glue should be more than adequate to hold everything secure.

Related info

Types of adhesive (see page 220)
Safer woodwork (see page 22)

Removing excess glue

Water-based glues – such as white and yellow glue – are straightforward to clean up, which is one reason for their popularity in general woodworking. Excess glue can be scraped away or removed with a damp rag while still wet. If the glue has dried before being cleaned up, then it can be removed with a sharp chisel or sanded away. Many non-water-based glues are difficult to clean up while still wet – read the manufacturer's recommendations for best results.

adhesives and assembly

types of adhesive

It is important to choose the most appropriate adhesive for any given task, since there is a wide range of different types. Some can be used for more than one purpose or material; others are more suitable for specialist tasks. Important considerations include the setting time, water resistance, bond strength, gap filling capability, ease of application and shelf life.

Glue storage

Many glues will be ruined if allowed to freeze. If your workshop is heated then this should not be a problem, but if it regularly freezes during winter then it is worth making a glue tote to carry your glue to and from the workshop as and when needed. Have a dedicated area of your workshop where you store glue. By doing this you can not only keep an eye on stock levels but it will also help to have all your gluing accessories in one easy-to-get-to place for those times during a glue-up when the pressure is on. Flammable products should be stored in a fire safe metal cabinet.

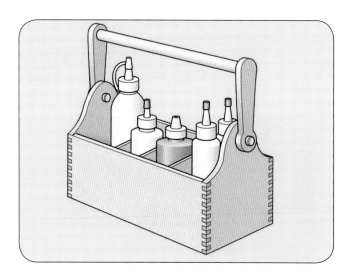

Animal glues

Animal glues were traditionally used by furniture makers for gluing joints together, but the performance of more modern adhesives – such as PVA – has relegated them to specialist tasks. These include small veneering and restoration jobs, where they are ideal because of their thermoplastic properties – they soften when heated. They are made by boiling bone and sinew for long periods of time; hide, fish and rabbit skin glues are all types of animal glue. Traditionally these glues were supplied in granules, which needed to be soaked in water overnight prior to being heated in a double skinned glue pot to prevent the glue from boiling. Some modern animal glues are now available in liquid form for convenience, but the shelf life of these is short so this type of adhesive should be purchased on a need-to-use basis.

+ Excellent for small repair and veneer jobs because of thermoplastic properties

– Short shelf life
– Tend to break down faster with age in comparison to modern adhesives
– Temperature dependent

Related info

Workshop storage (see page 18)
Safer woodwork (see page 22)

Planning ahead

Animal glues have a short shelf life so buy in small quantities as and when needed.

PVA

PVA glue – also known as white glue – is one of the most universally used adhesives in woodworking. It is cheap and versatile, while providing excellent strength for woodworking joints as well as good open times, so allowing enough time to assemble complex projects. When dry it turns translucent and can be awkward to sand, so it is best to clean up any squeeze-out with a damp cloth prior to the glue setting. PVA is supplied as a premixed emulsion of polyvinyl acetate suspended in water that sets when the water evaporates, so no additional mixing is required prior to use. Because some of the water is absorbed by the wood during the drying process mitre ends should be sized – coated with a thin layer of glue – and allowed to dry prior to being glued together, to achieve a stronger joint in its most basic form PVA provides no water resistance but it can be purchased in water resistant formulations. Aliphatic resin glue – also known as yellow glue – is a form of PVA that offers increased strength, faster setting times and sands easier than standard white PVA. Yellow glue is also available with an increased water and heat resistance and is a very popular all-round choice for woodworkers. Both white and yellow glues have very long shelf lives, but should not be allowed to freeze during winter.

+ Excellent all-round woodworking glue
+ Can be purchased with varying degrees of water resistance and open time
+ Very strong and reliable
+ Easy to clean up while wet
+ Good gap filling qualities up to 1mm (1/32in)
+ Long shelf life

– Not waterproof

Hot melt glue

Hot melt glue can either be purchased in sticks for use in a hot melt glue gun, or in sheet form for laying veneers. If using hot melt sheets, the heat from a domestic iron is used to soften and activate the adhesive after it has been laid between veneer and substrate. Hot melt glue is temperature dependent – although its thermoplastic properties allow it to be reheated and manipulated as desired. In stick form it is handy to have around the workshop for general repair jobs, quick jig making and the construction of scale models before embarking on the real thing, but otherwise it is not much use for woodworking and certainly not recommended for gluing joints together.

+ Great for model making and quick jig making
+ Can be used in sheet form for laying veneers
+ Convenient and quick

– Limited use in woodworking

Choosing the right PVA

Do not store PVA in freezing conditions. Most aliphatic resin PVA glues dry a yellow colour, but darker formulations are available for use with darker woods if gap-filling properties are required.

Small projects

Hot melt glue is excellent for jig and model making, especially when used with a dedicated glue gun – keep a stock to hand for all those little jobs around the workshop.

Polyurethane glues

Polyurethane glues are strong, versatile and waterproof. They are particularly useful when gluing long grain to end grain joints because – unlike water-based glues – they expand as they cure, which has the effect of driving the adhesive into the end grain to produce a stronger bond. The gap filling property of these glues is very good, but small bubbles are produced during the drying stage that are revealed during sanding and leave a rough surface that may require further treatment prior to finishing. Polyurethane glues are difficult to remove when wet so it is best not to attempt any clean up until the glue is fully dry. Disposable gloves should be worn during application because the glue can be very difficult to remove when it comes into contact with skin.

+ Excellent gap filling qualities
+ Very good for gluing end grain to long grain
+ No mixing required prior to use
+ High degree of water resistance

– Difficult to clean up while wet

Urea formaldehyde glues

Urea formaldehyde glue is a two-part adhesive usually supplied in powder form, but sometimes with a liquid catalyst. The two parts are mixed with water and then applied to both surfaces, before they are brought together and cramped (clamped). The glue cures by chemical reaction, forming a very strong bond that features high water resistance and good gap filling qualities. Care should be taken when handling urea formaldehyde – ensure the area is well ventilated and wear safety glasses and protective gloves during use.

+ High performance glue
+ Good gap filling properties
+ High water resistance
+ Long shelf life

– Potentially hazardous to health; safety precautions need to be taken during use

Resorcinol resin glues

Resorcinol resin glues perform much like urea formaldehyde glues, but are completely waterproof. They are supplied in two parts – either a resin with a liquid hardener or a two-part powder mix. When the glue is mixed it needs to be applied to all mating surfaces before they are cramped (clamped) and left to dry. Resorcinol glues are temperature dependent, with workshop conditions ideally being 15°C (59°F) or above. Ensure the area is well ventilated and wear safety glasses and protective gloves during use.

+ Waterproof
+ Very strong bond
+ Good gap filling properties
+ Long shelf life

– Potentially hazardous to health; safety precautions need to be taken during use

Shelf life

Polyurethane glues are light brown in colour when dry and have a shelf life of approximately one year.

Waterproofing

While highly water resistant urea formaldehyde is not waterproof – if you need a fully waterproof adhesive use resorcinol-resin glue.

Glue colour

Resorcinol resin glues dry a colour close to mahogany.

Contact adhesive

Contact adhesive is usually latex based and is useful when gluing Formica, veneers and other thin sheet materials to a substrate material over a relatively large area. To form a bond contact adhesive is applied in a thin layer to both surfaces and then allowed to dry to the touch; next the two surfaces are brought together to form an instant bond. There are different formulations are available allowing differing repositioning times – water-based formulations are safer but usually take longer to cure. Most contact adhesive is flammable, so adequate ventilation is necessary during use.

+ Excellent for bonding thin sheet material to substrate material
+ Allows repositioning (time varies depending on formulation)

– Noxious fumes
– Flammable

Epoxy resin

Epoxy resin is a high performance two-part synthetic adhesive usually supplied as a resin with hardener, but is also available in putty form. Its uses are limited in woodworking largely due to its high cost and viscosity, but for specialist areas where diverse materials need to be bonded together or repaired it can be invaluable. Epoxy resin is waterproof, dries clear and is used extensively in the boat-building industry and for specialist laminations.

+ Very high performance
+ Waterproof

– Viscous so difficult to apply to joints
– Relatively expensive

Cyanoacrylate (superglue)

Superglues have a whole host of uses in the workshop. In terms of woodworking they are particularly useful when gluing small intricate joints, such as mitres, because of their extremely fast setting time and cured strength. They are also very useful for small repair jobs and when making custom cabinet fittings because of their ability to join a wide variety of materials. Cyanoacrylate glues are available in varying consistencies from a thin liquid to a thick gel. Aerosol accelerators are also available to speed up the curing process.

+ Strong and fast
+ Will glue a multitude of material types
+ Available in different consistencies

– Overrun tends to leave white marks
– Extra care needs to be taken when using, due to rapid bonding rate

Glue properties

Various formulations of contact adhesive are available with differing properties.

Storage needs

Epoxy resin is sensitive to temperature and humidity levels so a good workshop environment is essential for reliable results.

Skin care

Superglues will bond skin to virtually anything very rapidly so always wear a pair of disposable gloves.

shaping and bending

Being able to form wood into curved and bent components opens up a whole host of possibilities. Shaping solid wood with cutting tools tends to generate a lot of waste material but still has its place in many woodworking projects. More economical forms of bending – such as laminating, kerfing and steam bending – are very useful in some instances. This chapter guides you through the different methods of shaping and bending wood.

drawing curves

Regardless of whether you are shaping or bending wood, the first step in creating good-looking curves is to draw them. If you have created curves on a drawing board – or computer and can print them full size – then you can simply take a copy and use it as a template on your workpiece ready for cutting but, more often than not, curves need to be scaled up to full size before you can think about cutting wood.

One method of enlarging curves from a drawing is to make enlarged copies on a photocopier; sections of a curve can be built up this way and rejoined to give an accurate representation of the original. A more common and recommended method is to draw the curve full size by hand. Often a scaled drawing is only intended to give an approximation of a finished set of curves – by drawing them full size you can tweak them in as much detail as you like before committing to shaping the material.

Related info

Measuring and marking out (see page 33)
Workshop rods and templates (see page 156)
Jigsaws (see page 70)
Bandsaws (see page 94)
Safer woodwork (see page 22)

Drawing full-size curves

Use a flexible metal ruler or a narrow section of flexible plywood to create long even curves.

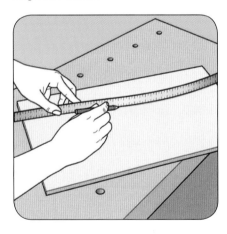

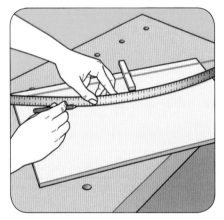

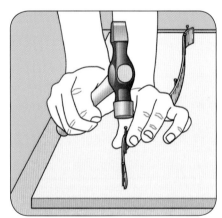

1 By drawing on manufactured board instead of paper you can create a rigid set of curves that can be turned into templates later if required. Hardboard is cheap and its smooth side is ideal for drawing on. If you need thicker material for making guide templates then 12mm (½in) MDF or plywood is very versatile, can be shaped accurately and also has enough thickness to run a bearing-guided router cutter against.

2 If you drill holes in either end of a flexible ruler, thread a length of string through and tie the ends of the string together, the ruler can be bent to a desired curve and held in position. This makes placement easy and allows repeat positioning for drawing. Use a batten to twist the string to fine-tune the curve.

3 Another method is to use panel pins at each end of the curve to hold both ends of the ruler in place, and then bend the ruler until the desired curve is reached. Keep hold of the ruler with one hand to hold it in place while drawing in the curve with the other hand.

Cutting templates

Using templates for curved work has several advantages. Once a template is created you can use it to copy a curve time and time again, while templates in at least 12mm (½in) thick material can be used with bearing-guided router cutters for easy repeat cuts with a router. Templates are used extensively for batch production (short production runs) of repetitive furniture components but are also useful when making single projects.

Making a template

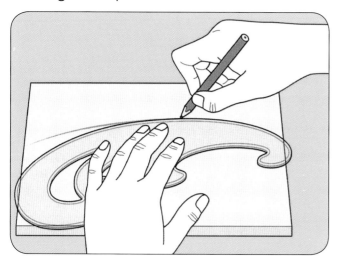

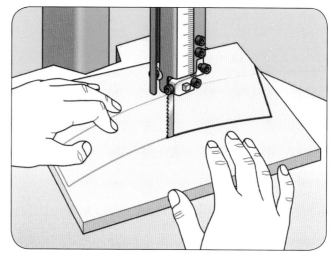

1 Choose a material for the template: 12mm (½in) thick MDF or plywood is a good choice for templates that will be used with bearing-guided router cutters. Clear acrylic or polycarbonate sheet is ideal if you intend using the template to draw around when shaping solid wood – being transparent, it allows you to see and select the grain beneath the template before drawing around it. A set of French curves is good for small details.

2 Cut out the template shape with a bandsaw or jigsaw. If cutting a plastic template, use a slow cutting speed to avoid the material melting and rejoining behind the cut. Cut slightly outside the line because both bandsaws and jigsaws will leave saw marks that will need cleaning up.

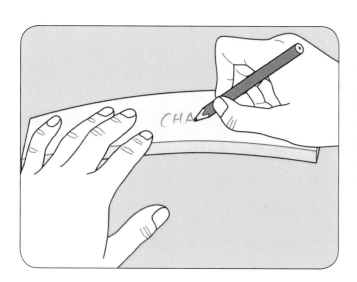

3 Remove saw marks with a combination of hand tools and medium to fine abrasive paper, being careful not to trim past the line. Plywood can be worked well with hand tools such as spokeshaves and planes but MDF is better shaped with abrasives, because it tends to dull cutting edges very quickly. Finish up by marking the template clearly with a component name and creation date – this is handy if you later decide to make an updated version but still want to retain the original for reference.

shaping board material

Templates are used extensively to shape manufactured boards. A bearing-guided router cutter can be used to follow the profile of a template, resulting in perfect copies time and time again.

Cutting template profiles with a router

Both profiled and straight cutters can be used in the router, depending on the desired result. If a template is fixed to the top side of the workpiece, a router table is used with the bearing guide at the top of the cutter when inverted in the router table. If the workpiece is cramped (clamped) to a template jig the workpiece and template position may be reversed, in which case the bearing needs to be at the opposite end of the router cutter closer to the surface of the table.

Related info

Making a template (see page 227)
Workshop rods and templates (see page 156)
Routers and cutters (see page 79)
Bandsaws (see page 94)
Safer woodwork (see page 22)

Cutting around the template

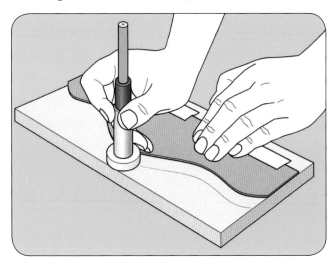

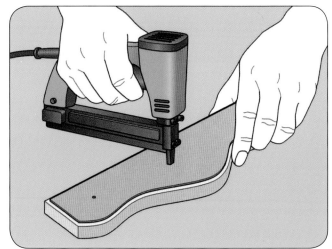

1 Draw around the template onto the material to be profiled and then use either a bandsaw, jigsaw or coping saw to cut close to the line. The aim here is to reduce the amount of material that the router cutter will need to remove, which results in an easier cut and a smoother finish.

2 Fix the template to the workpiece in one of several ways. If the template is to be fixed on top of the workpiece, panel pins or screws can be used to temporarily fix the two components together. Place fixings well away from the template edge to avoid contact with the router cutter – they should also be positioned so any subsequent holes will not be visible in the finished piece.

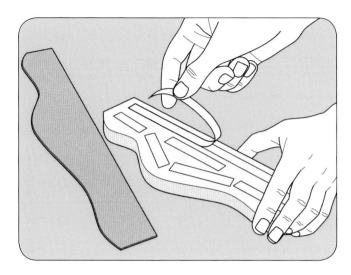

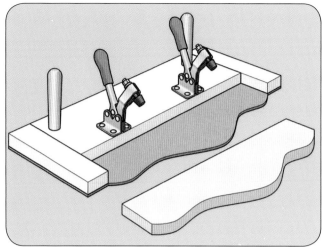

3 If holes are not desirable then industrial strength double-sided adhesive tape can be used to hold the template to the work. If tape is used test the bond has adequate strength before proceeding with the cut.

4 If the shape does not require the template to be followed all the way around its perimeter, cramps (clamps) can be used to hold template and workpiece together. Toggle cramps (clamps) are particularly useful because they can be fixed directly to the template to cramp (clamp) the workpiece in position.

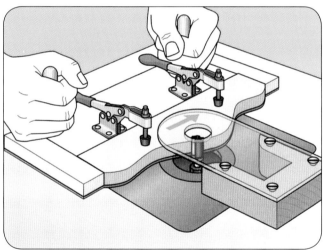

5 Start the cut part-way along the profile edge, away from any corners, or along a leading edge. Always feed the work against the rotation of the router cutter so that you are pushing against it. Feed the work at a steady rate keeping fingers well away from the rotating router cutter – always use an overhead guard.

Cutting template profiles with a bandsaw

Templates can also be used to cut curved profiles with a bandsaw. First fit your bandsaw with a workshop-made guide to follow the template – this is a basic device constructed from a narrow piece of wood with a rounded end so that it can follow a template profile. There should be a notch cut out for the blade so that the guide's curved edge can be set flush with the outer side of the bandsaw blade. A cutaway section on the underside of the guide provides clearance for the workpiece below the template.

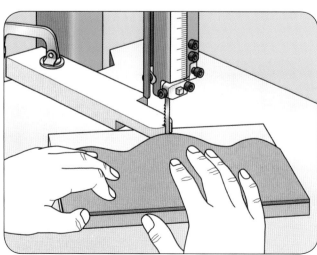

shaping solid wood

Shaping solid wood can be achieved with a wide variety of hand tools and power tools. For best results use a combination of both.

Simple shaping

Simple curves are those with curves running in only one direction at a time.

Related info

Measuring and marking out (see page 33)
Bandsaws (see page 94)
Spokeshaves (see page 51)
Sanders (see page 112)
Safer woodwork (see page 22)

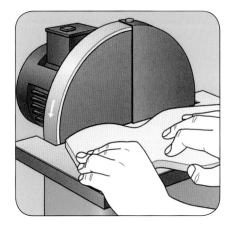

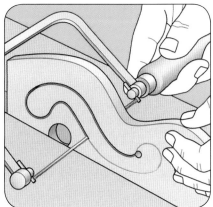

1 Start by cutting near to the curve line using a bandsaw fitted with a relatively narrow blade – the band saw excels at cutting curves and is usually the first choice for removing waste material. Curvature can be refined further with a hand tool; secure the workpiece in a vice so its curved edge is easy to work, and then use a spokeshave or compass plane to cut with the direction of the grain. For convex shapes work from both ends to avoid lifting the grain. Abrasive paper can also be used to achieve a good finish.

2 Alternatively you can use a static disc sander to work convex curves, making sure you use the section of the disc rotating towards the support table. A spindle/bobbin sander is ideal for refining internal curves – most spindle sanders come with a selection of different diameter interchangeable spindles for use on a wide range of internal radii. Good dust extraction and ventilation is essential when power sanding.

3 Using templates in conjunction with bearing-guided cutters – as outlined in the shaping board material section on pages 228–229 – also works well on wood of regular thickness.

However, for small intricate curvature on material up to 12mm (½in) in thickness there is no better tool than a scrollsaw. The scrollsaw is also one of the safest powered machines in the workshop because of its gentle oscillating blade action. Blades can be detached easily and threaded through holes drilled in the work if you need to start cutting from the centre of the workpiece instead of the edge.

Compound curves

Compound curves are created when two sets of curves flow in the same direction. A rear chair leg is a good example; the leg splays rearwards as well as sideways. Shown below are the steps for shaping a cabriole leg.

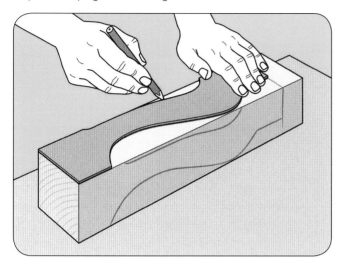

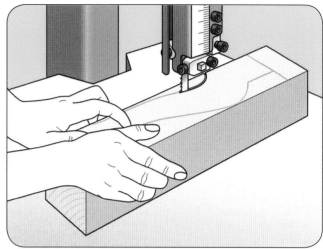

1 To cut a compound curve, first mark out the curvature on one plane and then rotate the wood 90 degrees and mark the second set of curves that will create the compound shape.

2 Select an appropriate size band saw blade for the thickness of wood and the curvature you intend to cut, and then start by cutting the first curve in one steady motion. If a second curve needs to be cut on the same face rotate the work so that the opposite side can be cut while using the same face as a reference on the bandsaw table.

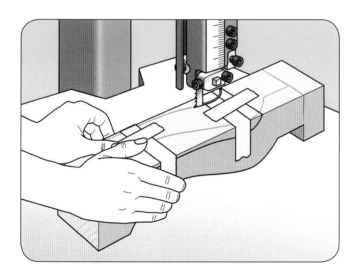

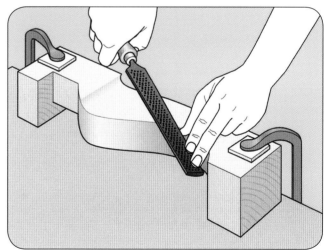

3 Use masking tape to temporarily fix the waste piece back in place. The work can now be rotated on the bandsaw table so that the second set of curves is visible on the taped waste sections.

4 Cut the second set of curves and then use a combination of compass plane, spokeshave, rasp and scraper to finish off the curves, working downhill with the direction of the grain.

shaping and bending

bending wood

Bending wood is one of those elements of woodworking that looks complicated but is in fact relatively simple. Exotic looking results can be achieved with a moderate amount of time and effort.

There are several ways of bending wood. The strongest method is to laminate thin layers of wood over a pre-shaped former; the only downside with this approach is that join lines between laminae remain visible – much like looking at the edge of a piece of plywood. Using a clear setting adhesive and cutting the laminae from the same piece of wood so that its grain pattern flows naturally can minimize this effect.

Kerf bending is another method of forming wood around a curve; it involves cutting deep grooves on the rear face of the curve to allow it to be bent round, leaving just a thin layer of unbroken wood on the visible front face. This method should be used selectively because it has some severe drawbacks. When bent, the cut grooves at the rear of the kerf will tend to project through to the front, leaving small visible facets on the curve. Kerf bent wood also needs a lot of reinforcement because it has no structural integrity of its own. Then there are the kerfs themselves, which are visible on three out of four faces. Even with these drawbacks kerf bending remains a very useful technique for certain situations.

Steam bending is a remarkable process where wood is impregnated with steam over a period of time until it becomes pliable. The wood is then strapped to a former and allowed to cool, so it sets and retains its bent form. Some degree of over-compensation is necessary when making the former because – unlike laminating – there will be a degree of spring-back in the curve that must be estimated and compensated for. If done well, steam bending can produce the most elegant and attractive of bent components.

Bent lamination

For this technique commercially available constructional veneers can be laminated together or wood can be cut into thin strips from solid stock to make individual laminate strips, also known as laminae. If cutting your own laminae use straight-grained wood that is free of knots and faults, because these will cause weakness and distortion in the finished laminate. Quarter sawn wood is ideal for this purpose because of its straight grain.

Related info

Veneers (see page 134)
Bandsaws (see page 94)
Cramps (clamps) (see page 36)
Safer woodwork (see page 22)

Preparing the wood

Given the choice of a bandsaw or a table saw to cut laminae, a bandsaw is preferable because, due to the thickness of the blade, the amount of material lost to waste is much less. There is also less risk of the strips being damaged during the cut.

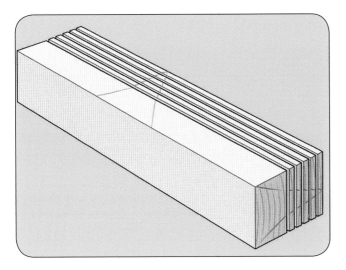

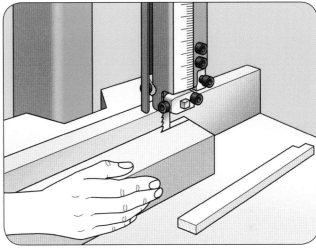

1 Before cutting all the strips experiment with different thicknesses of material to find the best for the radius you are trying to create. The thinner the material the easier it will bend, but the more laminae you will need to make up a given finished thickness. Start with a piece of squared up wood and mark a triangle on the side so that you can realign the grain pattern of the laminae when placing them on the former.

2 With the bandsaw fence set to slightly more than the finished strip thickness, cut the first strip. Plane the edge of the board square and repeat the process until you have enough strips to complete the job, with several spares in hand in case anything should go wrong. Use a thicknesser (thickness planer) to smooth the bandsawn faces of the laminae and bring them down to final thickness.

Making tight curves

If the geometry of your shape is particularly challenging soak the laminae in water, then cramp (clamp) them to the former and leave to dry before applying glue. This will partly preform the strips and allow tighter radii to be achieved.

Laminating with a two-part former

This process utilizes a male and female former to exert pressure, using cramps (clamps). Both the formers should have enough structural integrity to easily hold the laminae under pressure while the glue sets. Any weakness in the formers may cause distortion in the finished shape of the laminate, so build them either from solid material or with a strong and stable structure. Former surfaces should be sealed with sanding sealer and waxed to prevent glue from laminations bonding to the former.

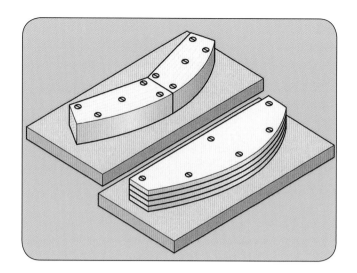

1 Male formers are easier to construct than female because the shape is more accessible with tools. Start by making the male former, either from solid wood or consecutive layers of manufactured board cut using a template. Material at the rear of the former should be approximately square to the former profile to enable an even cramping (clamping) pressure. Holes can be drilled into a solid former to insert cramp (clamp) heads.

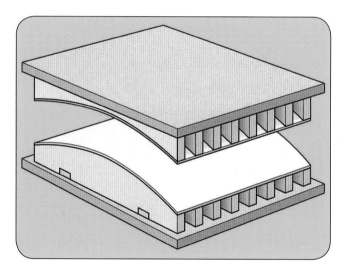

2 Formers for wide laminates – such as doors or panels – can be made by spacing the profiled components approximately 50–75mm (2–3in) apart and then covering them with a thin – 3mm (⅛in) – layer of plywood pinned and glued in place to complete the forming surface.

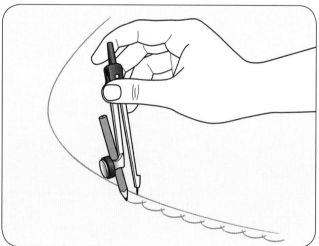

3 Female formers must have their profile offset by the thickness of the laminate being formed, otherwise the radii will not be correctly placed to apply equal pressure throughout the lamination. Scribe the male former profile onto a piece of manufactured board large enough to make a template for the female former. Use a compass set to the thickness of the lamination to draw a series of closely spaced arcs around the internal and external radii.

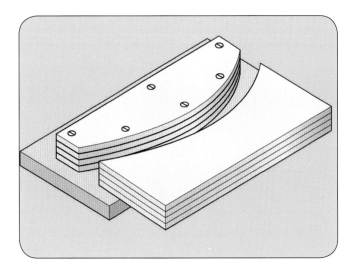

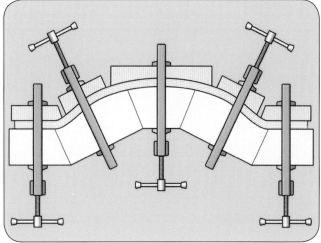

4 Join the high points of the arcs together to form the female profile. Cut out the profile to make a template and then construct the female former using the same method used to construct the male former.

5 When gluing laminae in place, use as many cramps (clamps) as practical, especially at the start and end of radii where cramping (clamping) pressure is most needed. Leave the laminate cramped (clamped) until the glue has completely set.

Using a single-part former with a vacuum bag press

The vacuum bag press has revolutionized laminating in recent years – and once you have used one it is very hard to go back to the traditional two-part former method of laminating. A vacuum bag press uses atmospheric pressure within a transparent bag to apply an even load across a male former, so no female former is required. This process opens up lots of possibilities, because the former build time is dramatically reduced. A male former can be built, tested and modified as required without having to repeat the process on the female half. And – because no female former is necessary – different thickness laminations can be made with ease.

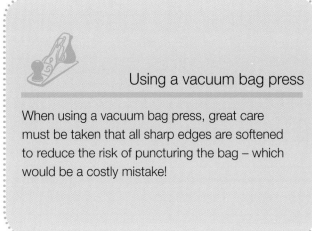

Using a vacuum bag press

When using a vacuum bag press, great care must be taken that all sharp edges are softened to reduce the risk of puncturing the bag – which would be a costly mistake!

Kerf bending

In kerf bending a series of controlled depth crosscuts are created to promote bendiness, while retaining the original thickness of the material. For best results kerf cuts should be equally spaced. The smoothest curves are achieved when the outer edges of the cuts are allowed to touch as the curve is bent or by using a male former with closely spaced cuts.

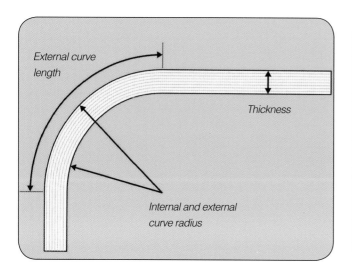

1 Use a full-size detail drawing to measure the external length of the curve, and then mark the start and end point onto the relevant section of the wood.

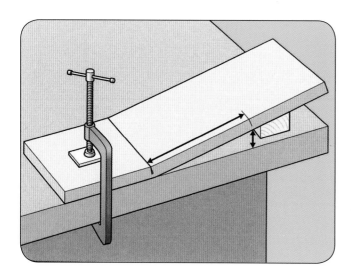

3 Cramp (clamp) the work securely to the bench at one end with the saw cut facing upwards and then bend the wood until the outer edges of the cut close. Place a wooden block underneath to hold the bend in position.

Calculating cuts for a 90-degree bend

This is a useful technique if you want the edges of the saw cuts to touch around the interior radii of a 90-degree curve.

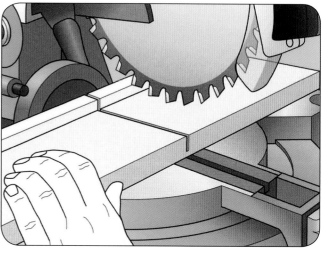

2 Make a controlled depth cut at the start point of the curve on the rear face – ideally using a mitre saw with a trenching facility or a radial arm saw, although a back saw can also be used to make the cut by hand. Leave at least 3mm (⅛in) of material intact on the front face of the curve.

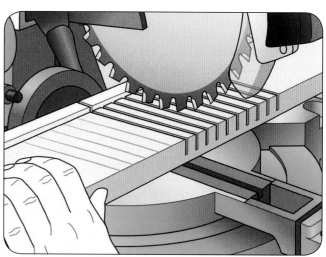

4 Measure the gap between the point opposite the end-of-curve mark made in step 1 and the bench top – this is the distance needed between cuts. Mark the cuts with a square and then cut them with the same saw that was used in step 2.

Making adjustments

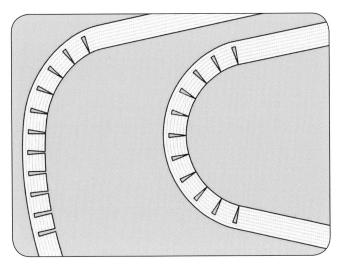

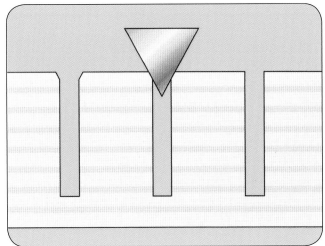

1 Bring the curve together to test how close it is to the desired result – a band cramp (clamp) can be used to hold the curve in position.

2 If the curve is too tight, insert slivers of veneer into each cut to open it out a little. If the curve is not tight enough, use a triangular file to remove an equal amount of material from each kerf edge, then retry. Repeat this process as necessary.

Fixing in position

A piece of veneer with its grain running in the same direction as the wood can be glued to the rear of the curve to lock it in place. Use wooden blocks to cramp (clamp) with an even pressure while holding the curve in place with a band cramp (clamp). A male former can also be used to hold the curve in position while it is under cramping (clamping) pressure.

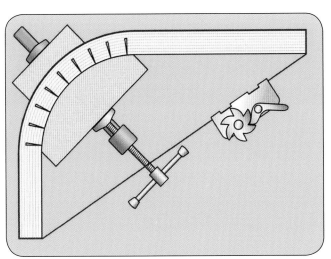

Irregular curves

For more irregular curves first create a male former and then experiment with different kerf spacing. Bear in mind that the closer the spacing between saw cuts, the smoother the curve will be. If the spacings are too large, then the curve will take on a faceted look and will need sanding to smooth it out. Use a male former to glue veneer tight to the back of the bend – ensuring the grain on the veneer runs in the same direction – to fix the curve in position.

Steam bending

Steam bending is a relatively simple process that can be achieved in most basic workshops. First the wood is heated in a steam chamber to approximately 1000°C (212°F), where it is held for at least 45 minutes for every 25mm (1in) of thickness. The wood is then removed from the chamber and bent around a male former, where it is left until cool. If only one component is required it should be left in the former for a period of two weeks; but if the former is needed for further bends, then a batten can be pinned across the ends of the bent profile once it is cool to hold it in shape. When the wood is released there will be a small degree of spring-back – how much is hard to estimate because it will differ from piece to piece. This spring-back effect will need to be factored in when making the former, and a degree of trial and error is likely before the right profile is achieved.

Warning

Keep the heat source well away from any flammable substances.

Always wear protective leather gloves and eyewear when steam bending.

Turn off and isolate the steam source before opening the steam chamber.

Make sure all elements are adequately vented to avoid excessive pressure build up.

The steam bender

1 A steam chamber to contain the wood for steaming can be made from marine grade plywood. It must be rebated (rabbeted), glued and screwed, with interior faces finished with exterior grade varnish to reduce moisture absorption. A simpler approach is to cut the steam chamber from a length of standard drainage pipe – this is well suited to steaming single pieces of wood at a time. If using drainage pipe it should be insulated on the outside to keep as much heat in the chamber as possible.

2 Place the steam chamber on trestles with the inlet end slightly elevated to allow water to be drained during the steaming process. Make end caps of marine plywood and fit them to both ends of the steam chamber, making sure that the cap on the outlet end has a bottom cutaway to allow water drainage. Place a bucket underneath to catch waste water.

3 Steam is supplied via a rubber hose inserted through a hole drilled at the centre of the inlet cap. Steam can be generated in a number of ways; a common method is to use a wallpaper stripper – by removing the steam plate the hose can be fed into the steam chamber to supply a continuous flow of steam. Another method is to use a hose attached to a modified metal drum, which is filled with water and placed on a portable gas or electric hot plate.

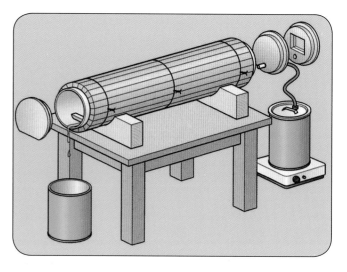

Woods suited to steaming

Some woods will steam bend more successfully than others because of their natural ability to become plastic and pliable when heated. Wood that has been air-dried rather than kiln-dried also retains more of its elasticity and so is better suited to bending. Some of the best woods for steaming include ash, beech, birch, elm, hickory, oak, spruce, teak, walnut and yew.

Making a former

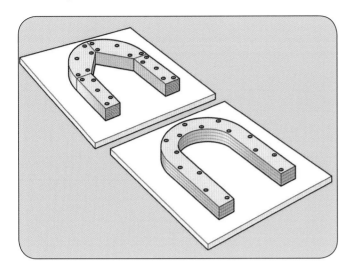

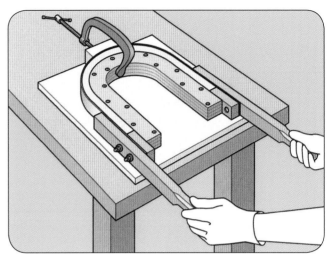

The bending process

When the wood is ready to come out of the steaming chamber it must be worked quickly. At this stage it is highly recommended that you have a helper nearby to place and tighten the cramps (clamps) when needed.

1 The former is made either from profiled sections of solid wood or laminated manufactured board, cut with the aid of a template and then fixed to a larger board that can be cramped (clamped) securely to a bench top. Large diameter holes are drilled through the former to allow cramps (clamps) to be used to secure the wood in place during the bending process.

1 Wearing thick leather gloves to protect your hands from the heat, turn off the steam source and take the wood out of the chamber. Place the wood between the stops at either end of the forming strap. Place one side on the former and have your helper fix the first cramp (clamp) firmly in place. Use the opposite handle on the strap to gently but rapidly form the wood around the former, with your helper adding cramps (clamps) as you go.

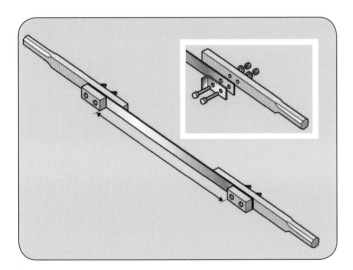

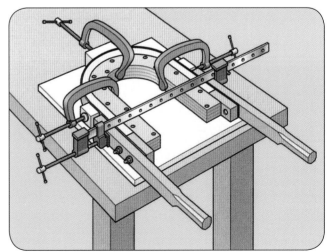

2 A forming strap will be needed to support the wood and stop its outer surface splitting as its fibres are stretched around the former. The fibres on the opposite side of the bend will be in compression and are much less problematic. The forming strap should have wooden end handles secured to a mild steel strap of 1.5rnrn ($\frac{1}{16}$in) thickness, and slightly wider than the section to be bent. If you are working with wood that reacts to steel – such as oak – then cover the strap in aluminium foil or polythene sheet. End stops are secured in place to stop the steamed wood slipping during bending.

2 Work your way around the curve, repeating the process until the entire section of wood is bent. Add more cramps (clamps), making sure that the wood is as tight to the former as possible. Leave it to cool completely.

shaping and bending

surface preparation and finishing

Good surface preparation and finishing is worth taking time over; if done well, all that hard work suddenly comes to life. Finishing can also be a daunting process because there are so many options and techniques available to the modern woodworker that knowing which path to take is not always straightforward. This chapter guides you through the process so that you can concentrate on what is important: completing your project.

surface preparation

Surface preparation is the key to good finishing. Without a good surface to work with achieving a high level finish is impossible, so putting the effort in at this stage will pay dividends later when you come to apply your chosen finish. The use of machines such as surface planers (jointers), thicknessers (thickness planers) and saws all leave telltale signs behind them; a seemingly flat surface from a thicknesser (thickness planer) may well be flat enough to work with and joint, but if you hold it up to the light you will notice lots of small scallops left by the rotating cutterblock. These types of marks will only be emphasized when any kind of finish is applied – when preparing a surface for finishing you should aim to leave no trace of the tools used in its preparation.

Related info

Planing and scraping
(see page 245)
Grain filling (see page 246)
Dust control (see page 20)
Safer woodwork
(see page 22)

Good surface preparation can be achieved in one of two ways: sanding surfaces by hand or machine can yield excellent results; planing and scraping surfaces requires only hand tools and can also deliver a first class finish. Which method you use comes down to personal preference, although the particular project at hand may have some bearing on your choice. Some woods plane better, while others require sanding to deal with tricky grain. You may prefer working with hand tools, in which case bench planes and scrapers will be your tools of choice, or perhaps you are attracted to the efficiency of power tools. Whichever route you decide on – and you should experiment with both – be assured that with moderate effort, both paths can result in the perfect surface for your finish.

sanding

This can be done by hand or machine. There are several types of power sander that are used extensively for surface preparation, although there are very few that can match the more traditional process of sanding by hand.

Power sanding

When power sanding, different machines are better for differing types of work. Belt sanders are the most aggressive of power sanders, used to flatten the surface if a large amount of material must be removed. They can dig into the workpiece easily, so care must be taken – fitting a sanding frame can help to keep the belt sander flat to the surface. Sanding across the grain will remove material very quickly, but always finish up by sanding with the grain.

Drum sanders are growing in popularity for the small workshop; they offer a highly efficient method of sanding surfaces flat and smooth. Different grades of abrasive grit (see box Abrasive paper grades on right) can be used to line the drum and the feed rate can be adjusted to suit the level of finish required.

Half- and third-sheet orbital sanders – also known as finishing sanders – are very useful for working large flat surfaces and removing material at a moderate rate without the risk of digging into the surface. Smaller palm sanders are ideal for smaller areas and lightly curved areas – their lighter weight makes them especially useful for working vertical surfaces.

Random orbit sanders provide the best surface of any of the orbital type sanders, because they feature an off-centre abrasive disc that rotates at random. This eccentric orbit action minimizes the circular scratching effect that is left on the surface of the wood by standard orbital sanders.

Hand sanding

Hand sanding is still used extensively alongside modern sanding power tools, particularly at the end of the sanding process to remove the last remaining signs that a machine has been used. For best results use a sanding block to keep the abrasive paper flat on the surface, and sand only with the grain to minimize the scratching effect of the abrasive paper. Sanding pads are now available with integral dust extraction to minimize dust pollution, along with its harmful effects.

Related info

Planing and scraping (see page 245)
Grain filling (see page 246)
Sanding sealers (see page 249)
Safer woodwork (see page 22)

Abrasive paper grades

80 grit Removes material quickly and flattens surfaces.

120 grit To remove any deep scratching or machine marks.

180 grit Will remove light scratching.

220 grit To achieve a good high quality finish.

320 grit For an exceptionally smooth surface.

400 grit Used to de-nib between coats of finish.

The sanding process

If using a belt or drum sander these machines are ideally used at the start of the sanding process to remove material quickly and flatten surfaces.

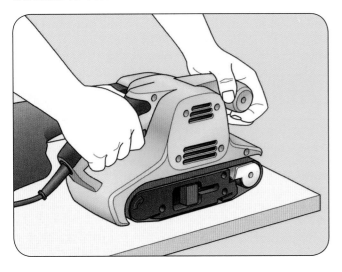

1 Start with 80 grit abrasive, and finish by sanding with the grain using 120 grit. If surfaces are already flat then this step is not necessary and you can start at step 2, using a random orbit sander or a standard orbital sander fitted with 120 grit abrasive.

2 Work the surface thoroughly and evenly, inspecting occasionally for marks. The aim is to remove any deep scratching left in step 1 or to remove any machine marks left during the process of planing and thicknessing the timber (lumber). Progress to 180 grit abrasive to remove any scratching left by the 120 grit abrasive, then move to 220 grit for a high quality finish.

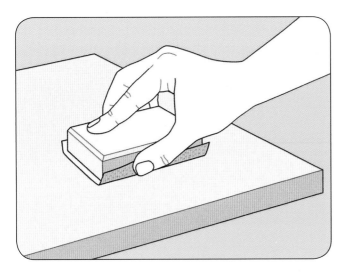

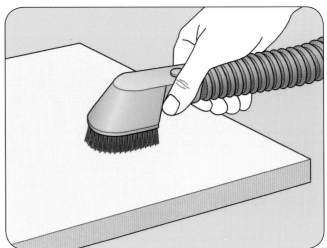

3 Next sand the surface by hand using a flat sanding block and 220 grit abrasive paper. You can progress to 320 grit for an exceptionally smooth surface, but there is no real advantage in going beyond this point. Remove the arris from exposed edges with a piece of 220 grit paper – softening edges like this can improve the look and feel of a project quite dramatically.

4 Finally use a vacuum cleaner with a soft brush attachment to remove dust from the pores of the wood. On open-grained woods, such as ash and oak, you will see a considerable difference in the appearance of the surface after performing this final step.

planing and scraping

Bench planes and scrapers are capable of producing the very finest of surfaces and, unlike abrasives, there is no need to work through different levels of abrasive grit. With a sharp iron and some practice you will be able to produce beautifully smooth surfaces ready for finishing quickly, and without the noise associated with power tools. Using edge tools to create final surfaces also eliminates the dull look sometimes associated with a sanded surface, which is caused by wood dust filling the pores.

The planing process

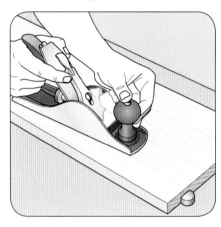

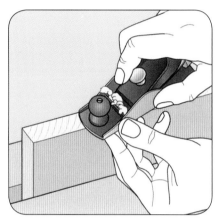

1 Use a finely set No.4 or No.4½ smoothing plane and work with the grain to avoid tear-out. Some woods respond well to being planed at an angle across their width, or even at right angles across their grain, but most woods respond best to being worked downhill in the direction of the grain. Experiment with a piece of scrap wood of the same type to see what approach works best.

2 A cabinet scraper or scraper plane can be used for tricky areas of grain, particularly when the use of a plane is causing tear-out no matter which direction you work. To use a cabinet scraper, place your thumbs behind the bottom edge and apply pressure so that the scraper bends and then work the wood with the direction of the grain. The area behind the cutting edge will become hot as you work – if it becomes too much put plasters on your thumbs to protect them from the build-up of heat. Cabinet scraper jigs are available to hold the scraper in its bent position while you work, which will virtually eliminate any pain caused by heat build-up.

3 A block plane with its mouth narrowly set is a great tool for removing the arris of exposed edges. First take several light cuts at 45 degrees then one at 22½ degrees on either side of the first cuts. Keep a count of how many cuts you make so you can repeat the same number on all arrises.

Related info

Sanding (see page 243)
Planes (see page 44)
Scrapers and spokeshaves
(see page 50)
Grain filling (see page 246)
Safer woodwork
(see page 22)

grain filing

Grain filling is an optional process that involves filling exposed pores to create a completely flat surface, which changes the look of the finished surface on open-grained woods. Woods such as oak, ash and elm benefit from grain filling because they feature relatively large open pores visible to the naked eye, but close-grain woods such as maple or beech have small pores that will not benefit from filling. If your aim is to create a high gloss finish, or a finish that appears to form a completely flat layer over the wood, then you should consider grain filling as part of your overall finishing process.

There are two types of grain filling: paste filling and direct filling. Paste filling uses a water- or oil-based filler that can be colour matched to the wood. Direct filling is achieved by building up successive layers of finish – usually the same material as the topcoat – until the pores are saturated. Direct filling has the advantage of being transparent, but it takes longer to produce because it is built up in multiple layers while most paste fillers require just one layer.

Pre-coloured paste fillers are available to create contrast and interesting effects. Liming wax is a good example; this is traditionally used on native hardwoods such as oak in domestic furniture. Before applying grain filler consider applying a sealing layer of blonde de-waxed shellac (see Applying sanding sealer on page 249), which will increase contrast by fixing the colour of the wood surface while leaving the larger pores of the wood open ready for filling.

Mixing your own grain filler

An off-the-shelf neutral colour grain filling product can be mixed with additional ingredients to colour match it to the wood. In these steps Van Dyke brown is being used to match mahogany, but substitute the relevant pigment to colour match any other wood.

Related info
Sanding (see page 243)
Planing and scraping (see page 245)
Sanding sealers (see page 249)

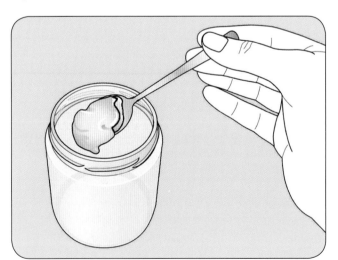

1 Place a scoop of grain filler in a glass container.

2 Add white spirit (mineral spirits) to approximately half the volume of filler then mix to a fine paste – this thins the filler and makes it easier to apply.

3 Add a small amount of Van Dyke brown pigment and mix well. You can always add more pigment so it is best to add too little rather than too much.

4 Apply a small amount of filler to a piece of scrap wood to test the colour match, and then add more pigment if necessary.

Applying grain filler

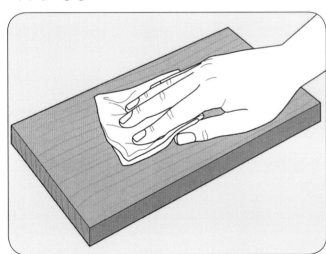

1 Apply grain filler sparingly with a rag across the grain. Before the filler has a chance to dry, wipe off any excess material and then set aside to dry thoroughly.

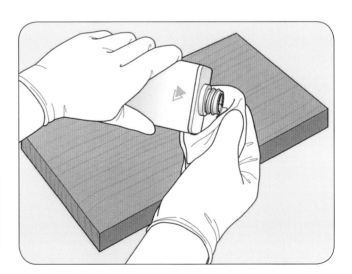

2 Remove any filler residue left on the surface with a sanding block fitted with fine abrasive paper. Apply a thin barrier layer of shellac – ideally blonde de-waxed shellac because of its transparent nature. This will aid adhesion of the topcoat and reduce the risk of any colour bleeding through from the grain filler.

Plaster of Paris

The use of plaster of Paris to grain fill – especially when French polishing – was a traditional method used in England. Modern off-the-shelf fillers have diminished its use somewhat these days, but it still has a place in some situations.

1 Scoop a portion of plaster of Paris into a glass container.

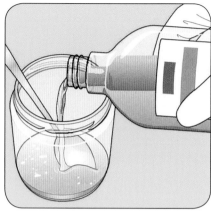

2 Pour methylated spirits (denatured alcohol) in slowly, mixing continually, to achieve a fine paste.

3 Apply the mix to the wood across the grain, using a rag. When the surface becomes dry to the touch remove any excess material with a clean damp cloth, and then allow to dry fully overnight.

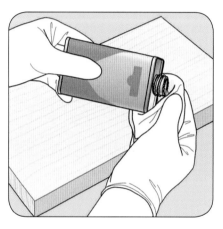

4 Use a sanding block fitted with 220 grit abrasive to sand flat. An opaque residue will be left on the surface, which will turn transparent if you apply a thin layer of boiled linseed oil.

5 When the boiled linseed oil is thoroughly dry, a barrier layer of blonde de-waxed shellac can be added prior to applying the topcoat.

Warning

Plaster of Paris contains crystalline silica – wear gloves and respiratory equipment and work in a well ventilated area.

sanding sealers

Sanding sealer is used either to seal the surface of wood prior to applying a topcoat of finish, or to provide a barrier layer between finishing materials that might otherwise react when layered directly together. Sanding sealers penetrate wood effectively and contain special compounds that aid sanding – they make it very easy and quick to de-nib (the process of sanding hardened areas of raised grain with fine abrasive paper) once dry. This is particularly useful because the first coat of finish always results in encapsulated areas of raised grain that need to be de-nibbed before proceeding to the next coat.

Sanding sealer is available with different solvent bases – acrylic, cellulose, shellac or spirit – which is useful when creating barrier layers between different finishing materials. For example, a water-based stain would need sealing prior to applying an acrylic lacquer because both products use wat er as a solvent so there would be a high risk of the stain bleeding through to the lacquer, especially if brushed on. A layer of shellac, cellulose or spirit-based sanding sealer would prevent this and provide a good smooth surface for the lacquer topcoat. Having different solvent bases means that sanding sealers can be used with virtually any combination of finishes.

Related info

Sanding (see page 243)
Waxes (see page 251)

Applying sanding sealer

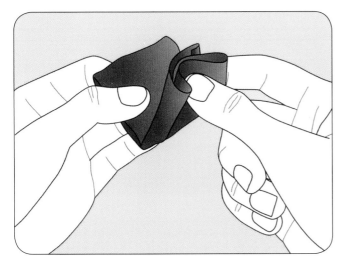

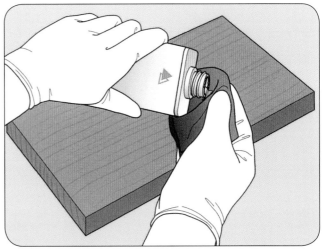

1 Apply sanding sealer with a wad made from cloth, or with a brush. A wad is recommended because it can be used to work the sealer into the wood in a circular motion as it is applied. To make a wad, first cut a section of lint-free material to approximate size and then fold it over itself several times, holding it together with your fingers.

2 Wearing protective gloves hold the wad over the neck of the sanding sealer bottle and gently rotate to transfer a small amount of sealer to the wad.

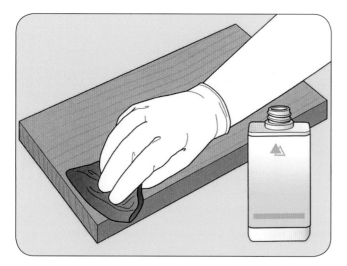

3 Apply the sealer to the wood in a small circular motion, starting at one corner and working outwards. For best results retain a wet edge by working at a pace that allows you to overlap the sealer before it dries.

4 Once applied sanding sealer dries quickly. Allow it to dry thoroughly, then de-nib with 320 to 400 grit abrasive paper. One layer of sealer is usually enough. Use a vacuum cleaner fitted with a soft brush attachment to remove any dust before applying the next layer of finish.

Sanding sealer and wax finish

One of the simplest and most rewarding finishes for wood consists of a base layer of sanding sealer followed by several topcoats of beeswax-based polish. This simple finishing recipe offers a natural looking finish without the high-build, high-gloss characteristics that modern lacquers often provide. For best results use on close-grained hardwoods such as maple and cherry. It is not suitable for areas of high wear or humidity, such as worktops, bathrooms or outdoor projects.

Using gloves

Finishing is a messy business – keep a stock of disposable gloves handy for all your finishing tasks.

waxes

Off-the-shelf waxes have grown in popularity in recent years and premixed waxes are available as pastes, liquids and sprays in a huge variety of colours and containing both natural and synthetic ingredients. Waxes are easy to apply, requiring just a little time and effort with no large learning curve necessary. They also result in some of the nicest surfaces, especially when combined with sanding sealers and oils.

Applying wax

The following method of applying wax will work well with most commercially available wax products. For best results seal the surface first with a layer of sanding sealer.

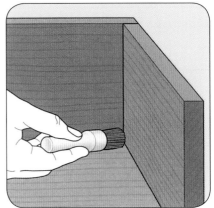

1 Apply a moderate amount of paste or liquid wax to a clean lint-free rag, then rub into the surface of the wood working with the direction of the grain to create a thin layer. Most spray waxes require application directly to the wood.

2 For hard-to-get-to areas – such as internal corners and details – use a dedicated wax brush to work the wax in. Avoid any excessive build-up of wax.

Related info

Sanding sealers
(see page 249)
Oils (see page 254)
Colouring (see page 257)

3 Allow the wax to partially dry – this usually takes several minutes but refer to the manufacturer's instructions for correct timing – then use a wax polishing brush or a clean soft rag to buff the surface to a sheen.

4 Set aside for two to three hours then reapply as desired. Reapply a fresh coat twice a year – over time a lovely patina will build up if wax is applied fairly regularly.

Making your own wax paste

Some of the best wax pastes are made from traditional naturally occurring ingredients. They are easy to make and provide immense satisfaction when applied.

Basic beeswax paste

This is a basic wax paste and is very easy to make. Try varying the amount of pure turpentine for a thinner or thicker paste.

Ingredients

50g (2oz) beeswax

120ml (4fl oz) pure turpentine

1 Grate the beeswax into a container. Measure out the pure turpentine before setting it aside. Turpentine is flammable so keep it well away from any direct sources of heat.

2 Bring a saucepan full of water to the boil and then place it on a hot plate. Pre-warm a heat resistant mixing bowl over the saucepan of hot water and then – working in a well ventilated area away from any direct heat source – add the beeswax and turpentine to the dry, warm bowl and stir gently.

3 When the wax has completely dissolved decant the mixture into an appropriate storage jar, cover and set aside for several hours to set. When the wax has set it is ready to use.

Ongoing care

Keep your furniture projects fresh and vibrant by applying a layer of wax every few months. For best results avoid furniture polishes containing silicone and stick to products containing beeswax and turpentine.

Hard-wearing wax for furniture

This recipe is a little more complex but produces a much harder wearing wax. Carnauba wax – used in many commercially available products such as shoe and car wax polishes – is very hard-wearing. Paraffin wax provides a good cost effective filling material to bulk out the wax alongside the beeswax.

Ingredients

130ml (4½fl oz) pure turpentine

1 tablespoon of carnauba wax

25g (1oz) paraffin wax

25g (1oz) beeswax

1 Measure out the ingredients and set them aside well away from any direct heat sources. Bring a saucepan of water to the boil, then move it to a hot plate.

2 Pre-warm a heat resistant glass bowl over the saucepan of hot water and then add approximately half the pure turpentine to bring it up to temperature. Carnauba wax has one of the highest melting points of any naturally occurring wax, so add this next and stir until completely dissolved. If it fails to dissolve fully, replace the water in the pan with boiling water from a kettle.

3 Add the remainder of the pure turpentine followed by the paraffin wax and beeswax. Continue stirring until all the ingredients have melted and a nice even consistency has been achieved. Decant into an appropriate storage jar, cover and put aside to set.

surface preparation and finishing

oils

Oil finishes can produce stunning results, especially on dark hardwoods. They can also be used in conjunction with other finishes to give wood grain a little extra visual pop, prior to applying a topcoat. Most oils are relatively cheap, versatile in use and easy to apply. Pure oils can be used neat or thinned for extra penetration into the wood, while ready-made off-the-shelf oil-based products offer a wide variety of finishing options.

Types of oil finish

Of all the oil products available to the woodworker, only a small number are classed as true oils. Linseed oil and tung oil are the most common true oils used for wood finishing, primarily because of their ability to dry without adding drying agents. Semi-drying true oils – such as safflower and soya – and non-drying true oils – such as coconut – are not applied directly to wood but these oils are found in commercially available varnishes with added drying agents.

Danish oil, finishing oil and teak oil products are not true oils because they consist of several ingredients – in fact these products are closer to varnishes than oils because they form a thin hard film on the surface of the wood. However, because they do still contain oil and are thin enough to penetrate deep into the wood they still fall under the oil category.

Linseed oil

Available as raw linseed oil or boiled linseed oil, this well established true oil is extensively used in wood finishing. Traditionally linseed oil was actually boiled and then allowed to cool to speed up the drying process after application, but modern production methods tend to use added chemical dryers in place of boiling even though boiled linseed oil retains its traditional name. Raw linseed oil has a longer drying time so – since boiled linseed oil is readily available – its popularity has decreased significantly over recent years.

Food safe oils

Because of its natural water repellent qualities tung oil is a traditional finish for kitchen worktops and utensils. In its true oil state it is non-toxic, but manufacturers often add metallic dryers to pure oils to speed up the drying process so when looking for an appropriate finish for kitchen accessories and utensils always check that what you plan to use is safe. Some oil products are sold specifically for this purpose – salad bowl oil is one example, and olive oil can also be used as a food safe finish.

Related info

Waxes (see page 251)
Colouring (see page 257)

Warning

Oily rags will generate heat and they can spontaneously combust – always douse them in water and place in an outdoor bin straight after use.

Applying true oils

True oils – boiled linseed oil or tung oil – should be applied in several stages.

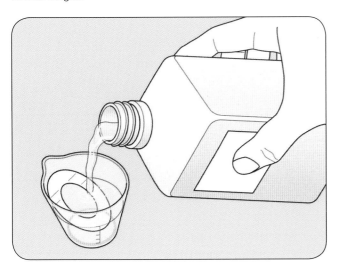

1 Thin the oil to a 50/50 mix, using either pure turpentine or white spirit (mineral spirits).

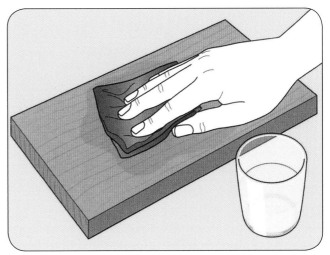

2 Apply generously to exterior surfaces with a lint-free rag or brush and allow to soak in before wiping off any excess oil with a clean rag. Avoid applying to interior surfaces because true oil tends to go rancid in confined spaces.

3 After 24 hours a 75/25 mix of oil/thinner can be applied as in step 2. After another 24 hours apply pure oil thinly – it can be heated slightly in a heat resistant glass bowl over a saucepan of hot water, which will thin it and make it easier to apply.

Getting the best oil finish

For an optimum oil finish apply one coat every day for a week, then once a week for a month, then once a month for a year, followed by annual coats to maintain the finish.

Off-the-shelf oil products do not usually need thinning, but because recommended methods of application will differ from product to product it is always worth reading the manufacturer's guidelines on how to apply their specific product.

surface preparation and finishing

Making your own oil-based polish

Boiled linseed oil and tung oil mix very well with traditional waxes and this can result in some stunning surface finishes, particularly on medium to dark hardwoods such as walnut, mahogany or brown oak.

Linseed oil and wax polish

This easy and effective wax has carnauba wax added to make the finish more resilient, but you can omit this if you prefer.

Ingredients

60ml (2fl oz) pure turpentine
1 tablespoon of carnauba wax (optional)
100ml (3½fl oz) boiled linseed oil
50g (2oz) beeswax

1 Bring a saucepan of water to the boil, then move it to a hotplate. Pre-warm a heat resistant glass bowl over the saucepan full of hot water. Add approximately half the pure turpentine then the carnauba wax and stir until dissolved. If the carnauba wax fails to fully dissolve replace the water in the saucepan with boiling water from a kettle.

2 Add the remaining pure turpentine along with the boiled linseed oil and continue to stir before adding the beeswax.

3 Continue stirring until all the ingredients are fully dissolved and an even consistency is reached. While still hot decant the mixture into an appropriate size storage jar, cover and set aside to set.

colouring

Wood can be coloured for many reasons. Often it is to give the impression that an item is made from a more exotic type of wood – making inexpensive hardwood look like something luxurious and desirable, for example. Wood may also be coloured to even tone, equalizing the variance between heartwood and sapwood, or to achieve a more dramatic effect. Whatever the reason for colouring wood there are a number of ways to achieve the effect and a multitude of available products.

Some manufacturers combine colour with other finishing products – for example, coloured waxes and shellacs will allow you to colour wood while simultaneously applying layers of finish. This can be a useful approach if time is an issue and you just need to achieve an approximate colour change, but if you need more control you will need wood stains. There are two types in common use: pigment-based stains and dye-based stains.

Related info
Wood characteristics (see page 124)
Surface preparation (see page 242)
Waxes (see page 251)

Pigment-based stains

These consist of three main ingredients – pigment, binder and thinner – and contain inert particles that do not dissolve in liquid. The pigment is held in suspension by the binder, which allows it to stick to wood. Common binders include tung oil, boiled linseed oil, urethanes and acrylics. The binder generally determines the consumer classification, so products with tung oil or boiled linseed oil binders will be classed as oil-based while products with acrylic-based binders will be classed as water-based. The viscosity of the stain is controlled by the thinner, which also aids the drying process. The pigments used for colour may be natural or manufactured, Natural pigments are derived from the earth – ochres, siennas, and umbers – and they tend to be earthier in colour. The brighter coloured pigments – such as red, yellow and blue – are manufactured using a variety of chemicals.

Pigment-based stains can be used to increase contrast on open grained woods such as ash, oak and elm. Medium grained woods, like mahogany, take pigment stain well but because the grain is less exposed less contrast is produced. Close-grained woods, like sycamore and maple, do not take pigment-based stains well because the pigment is not absorbed and tends to sit on the surface. For these woods you can try building up multiple layers to increase the colour, although grain definition will soon start to look muddy so this approach is best avoided. Surface texture also affects how the pigment stain will take: if the wood is too smooth a pigment stain may not take well. In this instance the grain could be raised with a damp cloth before applying the stain, which roughens the surface slightly and allows the stain to stick.

Dye-based stains

Dye-based stains do not suffer from the same drawbacks because the dye is dissolved in liquid solvent to produce solutions with microscopic particles, much smaller than those found in pigment stains. This allows dye-based stains to penetrate wood much more effectively and evenly, resulting in a more pleasing colour on close-grained woods. Dye stains work well on open-grained woods as well, but – because of their even rate of absorption – do not generate the same level of contrast as pigment-based stains.

Dye-based stains are also far more transparent than pigment-based stains because of the smaller size of the colour particles, which makes them ideal for highly figured woods where it is desirable to retain as much definition in the grain as possible.

The main downside to dye-based stains is that they are nowhere near as light-fast as pigment-based stains – especially earth pigments – and will fade over time. The rate of fading is increased when placed in direct sunlight.

Making your own stains

There are numerous off-the-shelf stains available in stores but for best results it is worth making your own because you can experiment until you get just the right colour. Two recipes, for making your own pigment-based and dye-based stain, are given on pages 258–259.

Dye-based stain

Van Dyke brown crystals can be used to make a water-based dye stain for colouring wood any shade from a light yellow through to black.

Ingredients

Warm water

Van Dyke brown crystals

1 Add warm water to a mixing bowl and then add a small amount of Van Dyke brown and mix until dissolved.

2 Apply the solution to a piece of scrap wood with a lint-free rag to assess the colour. To lighten the dye add more water; to darken it add more crystals.

3 Once the wood is stained to the correct colour, seal it with a coat of blonde de-waxed shellac before applying your topcoat.

Achieving an even tone

Stains are often used to recolour wood but you can also use them to even out the tone between heartwood and sapwood. Always test on a piece scrap material first.

Pigment-based stain

This pigment stain is oil-based – you can add any colour, but the steps show it being made with burnt sienna pigment.

Ingredients

60ml (2fl oz) boiled linseed oil
Pigment of your choice
60ml (2fl oz) pure turpentine

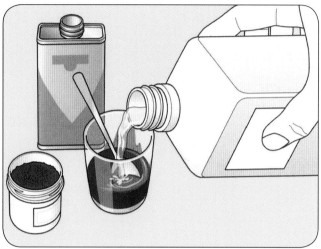

1   Add a small amount of boiled linseed or tung oil to a mixing bowl, then add a dash of pigment and mix with a teaspoon.

2   Add pure turpentine to the same proportion as the oil and mix thoroughly.

3   Test the stain on a scrap piece of wood. To darken, add more pigment. To lighten, add more oil and turpentine in equal measure.

surface preparation and finishing

french polishing

The traditional process of French polishing involves applying this fast drying polish with a rubber, or special brush, in a series of layers to build a high lustre, high quality finish. French polish can also be used in combination with other finishing materials such as stains, oils and waxes to produce some of the best looking finishes available for wood.

French polish is a solution of shellac dissolved in methylated spirits (denatured alcohol). Shellac – a substance secreted by the lac beetle – is collected and then processed into flakes. It can be bought either as flakes or as off-the-shelf premixed polish in a variety of colours – the most common of which are:

Garnet polish Dark brown, ideal for dark woods and restoration projects.

Button polish Light golden brown, also for use on restoration work and dark woods.

White polish Milky translucent, suitable for pale to mid-tone woods.

Pale or de-waxed polish Minimal colour, almost transparent, enhances the natural grain pattern of most woods.

Black polish Black, used in conjunction with ebony dye to ebonize (make it look like ebony) wood.

Related info

Surface preparation (see page 242)
Grain filling (see page 246)
Waxes (see page 251)
Oils (see page 254)
Colouring (see page 257)

French polishing tools

A traditional polishing rubber is made with a lint-free cotton or linen outer layer, which is wrapped around a cotton waste (or wadding) absorbent core that acts as a reservoir to store the polish during use. Alternatively a French polishing mop – a large round brush with very soft hair specifically for French polishing – can be used, but this is expensive so a rubber is preferable if you are new to French polishing.

Making a polishing rubber

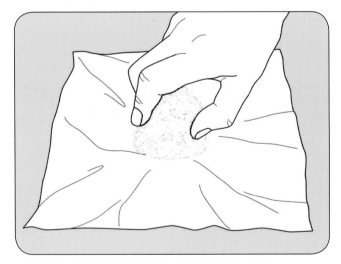

1 Cut a square section of lint-free cotton or linen and lay it flat. Scrunch up a moderate handful of wadding or cotton waste and place it in the centre of the material square.

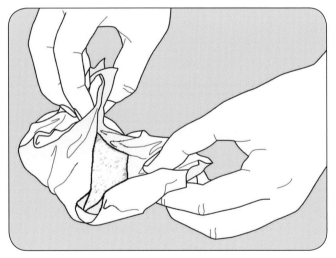

2 Bring all four corners of the material up and twist them together at the top to form an oval shaped pad.

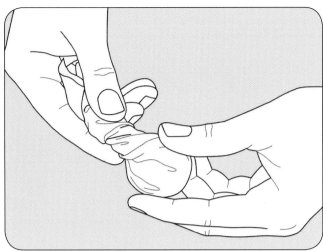

3 Make sure the base of the pad is flat and free of creases. You can vary the size of the rubber to match the project.

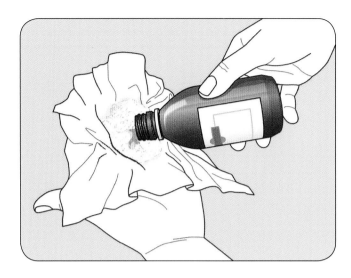

4 Charge the rubber with French polish – this is done by unwinding the outer fabric of the rubber to access the core and then pouring in a moderate amount of French polish.

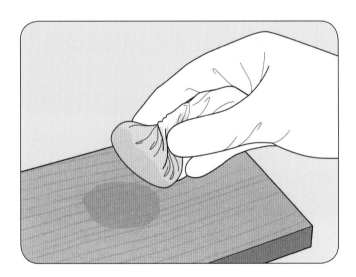

5 Twist the rubber together again and test it by dabbing it on some scrap material. Avoid applying French polish directly to the outside of the rubber, because the outer layer acts as a filter.

6 To increase the flow, tighten the winding at the back. Store the rubber in an airtight glass jar with a small amount of methylated spirits (denatured alcohol) to stop it drying out.

French polishing different woods

Medium to dark hardwoods often benefit from a layer of boiled linseed oil prior to applying French polish. This can enhance the grain and give the end result more visual depth and quality. Open-grained woods will benefit from grain filling beforehand; this can either be done with a grain filler paste or by building up successive layers of shellac until the grain is "full", then cutting back with fine abrasive before starting to French polish.

Applying French polish

French polish is built up in three stages.

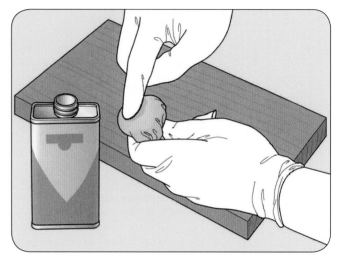

1 The first stage is called fading up and is repeated several times to seal the wood. Use the rubber to apply polish with the direction of the grain, making sure that each pass overlaps the previous one slightly to retain a wet edge. To avoid marking, bring the rubber into contact with the wood in a sweeping motion so that it is never stationary on the surface.

2 The second stage is called bodying in and involves building up the polish in thin layers. Apply the polish with a circular or figure-of-eight motion, being sure to keep a wet edge as you work. If the rubber becomes tacky, dab a little boiled linseed oil onto its surface – this acts as a lubricant to keep the polish flowing nicely. After 4–6 layers of polish set the work aside to harden for several hours, and then continue bodying in until you are happy with the depth of the finish.

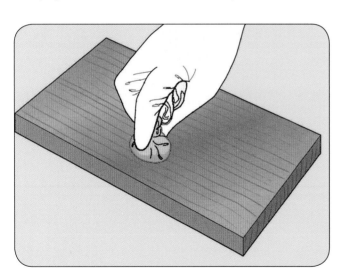

3 The final stage is called spiriting off and a new rubber will be needed. Charge the rubber with thinned French polish; two parts polish to one part methylated spirits (denatured alcohol). Apply the spiriting off layer with the same circular or figure-of-eight motion used when bodying in. Repeat this stage 2–3 times, then reload the rubber with a 50/50 mix of polish/methylated spirits (denatured alcohol) and continue for a further two coats, moving the rubber quickly and in the direction of the grain for a gloss finish.

Getting the best finish

For a high gloss finish cut the surface back slightly with 0000 grade steel wool, then apply burnishing cream and polish vigorously with a cotton rag until a high shine is achieved. To avoid a high gloss finish, gently cut the surface back with 0000 grade steel wool, then apply beeswax paste polish before buffing to a gentle sheen.

Making your own French polish

There are advantages to mixing your own French polish: it has a limited shelf life when premixed so storing shellac in flake form can be cheaper if you do not use French polish often, and you can make an exact colour by mixing different colour shellac flakes together.

French polish

Ingredients

500ml (17fl oz) methylated spirits (denatured alcohol)

250g (8oz) shellac flakes

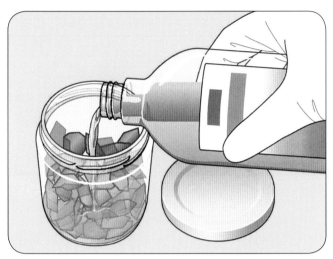

1 Put the shellac flakes in a glass jar with a liquid-tight lid and add the methylated spirits (denatured alcohol).

2 Shake the contents of the jar well and then set aside for 24 hours for the shellac to dissolve. Agitate the jar every few hours until the flakes fully dissolve.

Adjusting the viscosity

If you prefer a much thicker French polish, try doubling the amount of flakes to 500g (1lb 2oz) per 500ml (17fl oz) of methylated spirits (denatured alcohol).

lacquers and varnishes

Modern lacquers and varnishes have come a long way in recent years. Until recently the dividing line between lacquers and varnishes was clear: lacquers dried through evaporation of the solvent, while varnishes – containing a mix of oils, resins and solvent – dried through a combination of solvent evaporation and oxidation. Modern equivalents are so complex now that this dividing line has become blurred but manufacturers still refer to their finishes as either lacquers or varnishes to avoid confusion.

Oil varnishes

Modern oil varnishes contain synthetic resins – such as polyurethane, alkyd or phenolic – and dry by oxidation to form a hard-wearing non-reversible coating. White spirit (mineral spirits) is normally used as a thinner and is used to clean brushes. Quick drying oil-based varnishes are available for both interior and exterior use in a variety of finish options. The majority of liquid varnishes are designed to be applied by brush but many can also be sprayed.

+ Excellent surface protection
+ Easy to apply
+ Vast choice available

– Fumes given off during application, so requires a well ventilated space

Related info

Surface preparation (see page 242)
Safer woodwork (see page 22)

Acrylic finishes

The development of acrylic finishes is relatively recent when compared to more traditional lacquers and varnishes, and quality has been dramatically improved in recent years. Acrylic finishes include both lacquers and varnishes; lacquers are usually intended to be sprayed, while varnishes are primarily intended to be brushed on – although both types can usually be brushed or sprayed if thinned correctly. Since they are water-based acrylic finishes are some of the safest to use, and because they are not flammable they can be sprayed in a well ventilated area without using a spray booth.

+ Non-flammable
+ Can be sprayed in domestic workshop
+ Wide range of quick drying finishes available
+ Brushes can be cleaned in tap water

– Temperature and humidity need to be correct for thorough drying
– Surfaces can feel plastic

Effective varnishing

When applying varnish by brush, try thinning it to a 50/50 mix of varnish to solvent for the first couple of coats – this allows it to soak into the wood more effectively.

Two-pack lacquers

This category includes any lacquer that requires mixing with a catalyst prior to use. Most two-pack lacquers are very hard-wearing so are ideal for areas of high wear and tear. They provide a good level of protection against water and heat .

+ Very tough surface protection
+ Available in a huge range of surface finishes

– More complicated to apply than oil varnishes
– Most two-pack lacquers require fumes to be extracted for safe use

Pre-catalyzed cellulose lacquer

Cellulose lacquer has long been popular with furniture makers because it gives a good level of protection and dries very quickly. It dries by evaporation of solvent and this is reversible because it will soften with the application of cellulose thinners. Since it dries so fast this type of lacquer is only really suitable for spraying, and because it is highly flammable a spray booth is required to ensure that fumes are safely extracted.

+ Fast drying
+ Good quality finish with medium level of protection
+ Dries clear

– Highly flammable
– Only suitable for spraying with adequate fume extraction

hardware

Woodworkers have never had so much choice when it comes to hardware. Fixtures and fittings – including hinges, locks, catches and handles – are available to suit all sorts of traditional and contemporary styles, while knock-down fittings can provide convenience and ease of construction. This chapter looks at some of the common types of hardware to give you an idea of what is available, so you will know what to look for to carry out a particular function.

hinges

Hinges perform a variety of functions, from simply allowing a door to be opened and closed easily to providing integrated stops and soft close mechanisms. With so many types on the market you will be sure to find the correct one for your project.

Butt hinge

Traditionally used by cabinetmakers for a whole host of different jobs, the butt hinge is still one of the most popular hinges in use today because of its versatility. With so many different sizes available the butt hinge is ideal for a wide range of jobs, from hanging large cabinet doors to fitting jewellery box lids. Good quality butt hinges are made from solid brass.

Lift off hinge

The lift off hinge is used when hinged doors or panels may need to be removed on occasion – rather than unscrewing the hinge, the door can just be lifted up and off. Lift off hinges are fitted in a similar way to butt hinges.

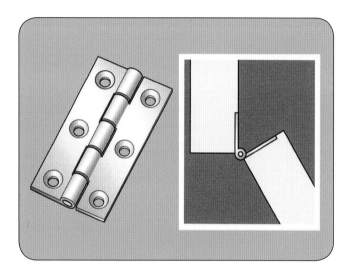

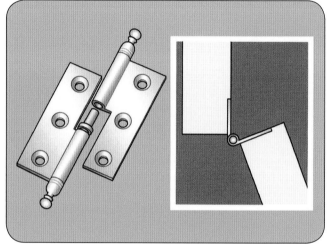

Related info

Project design (see pages 146–159)
Locks and catches (see page 271)

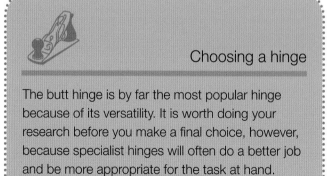

Choosing a hinge

The butt hinge is by far the most popular hinge because of its versatility. It is worth doing your research before you make a final choice, however, because specialist hinges will often do a better job and be more appropriate for the task at hand.

Flush hinge

A cheaper version of the butt hinge, usually made from pressed steel but coloured to look like brass. It is easier to fit because flush hinges do not require recesses to be cut in the wood of the door and the frame.

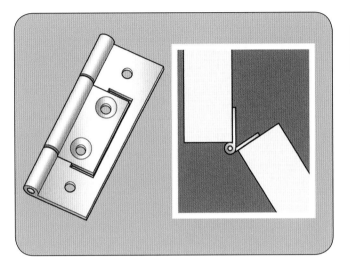

Concealed cabinet hinge

Used extensively in knock-down furniture and kitchen cabinets, the concealed cabinet hinge allows doors to be adjusted both vertically and horizontally so several doors in a row can be lined up easily. Doors also open in line with the cabinet sides, so they can be fitted in close proximity to each other. Fitting is straightforward and usually involves drilling a hole to sink a portion of the hinge, and then fixing both sides in place with screw fixings.

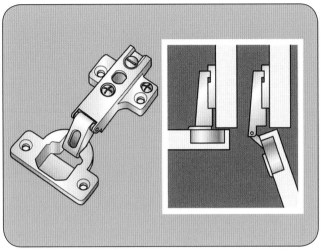

Cranked hinge

This hinge improves access by carrying the door leaf clear of the effective opening width and is often used to fit cabinet doors in high-end furniture work.

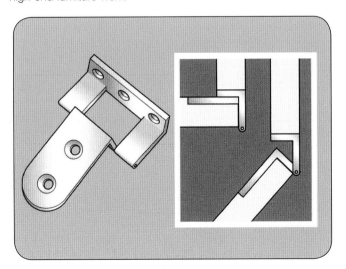

Back flap hinge

The "wings" of this hinge are wider than those of a butt hinge; it is used primarily to attach bureau fall flaps or to fit lifting leaves to tables.

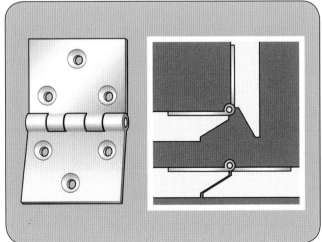

hardware

Pivot hinge

Pivot hinges are available in straight or L-shape profiles and are used to fit doors in fine cabinetwork. They are fitted at the top and bottom edges of doors and provide a well concealed hinging solution.

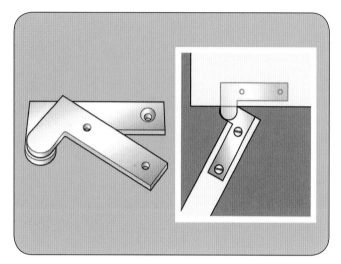

Table hinge

An offset hinge much like the back flap hinge, but used to hinge rule joints when making fold-down table flaps.

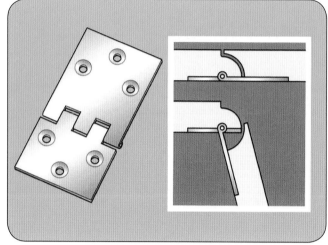

Piano hinge

Available in various off-the-shelf lengths, the piano hinge is cut to length to fit the project and is used where exceptionally strong hinges are needed. Cabinets with inside door storage, which cause excess strain to be placed on the hinge, benefit from having piano hinges fitted.

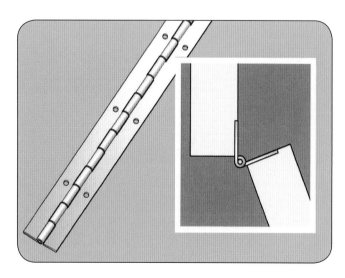

Quadrant hinge

Quadrant hinges are used for fitting box lids – they are more expensive than butt hinges but have a built-in stop mechanism for keeping the box lid open.

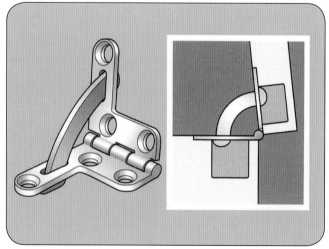

locks and catches

Well made locks can add elegance to jewellery boxes and furniture, as well as providing a degree of security. The types of locks featured here are designed to preserve privacy rather than provide high security. Catches can add the finishing touch to cabinet doors by providing positive stops that are satisfying to use.

Cabinet and box lock

The traditional cabinet lock is used to secure cabinet doors and drawers. The main housing of the lock is recessed into the rear of the door and then an escutcheon is added around the keyhole on the visible outside face. Box locks are similar to cabinet locks, but feature a hooked locking pin – in place of the straight locking pin found on the cabinet lock – to allow the lock to retain hold of a box lid.

Cylinder lock

Cylinder locks are housed in holes drilled though the wood and are fixed with screws from the rear, leaving a small round face, complete with keyhole, exposed on the cabinet front. Sliding door locks are a form of cylinder lock that feature a retractable bolt, which locks into a socket on the rear sliding door. A key retracts the bolt to unlock the sliding door.

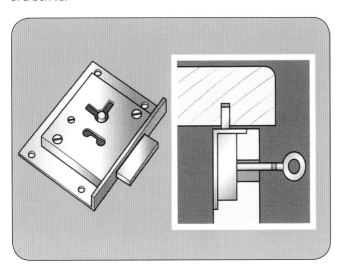

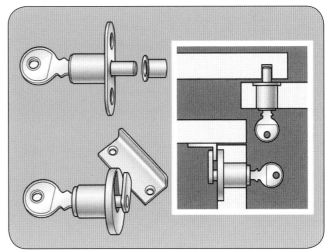

Related info

Project design (see pages 146–159)
Hinges (see page 268)

Lock or not?

Furniture locks tend to be designed more for privacy than security. If a thief wants to get into a piece of wooden furniture they can usually find a way – bear this in mind when deciding whether to fit a lock to your project.

hardware

Magnetic catch

Magnetic catches consist of a magnet housed in a plastic casing, which is fixed inside a cabinet, and an oval shaped metal striker plate that is fixed to the inside face of the door. When the door is closed the plate is attracted to the magnet, holding the door shut. These catches work well and are commonly used in knock-down furniture.

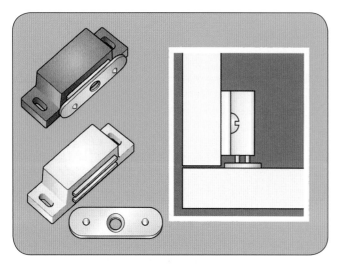

Ball catch

This traditional type of brass catch uses a spring-loaded ball bearing that is fixed in the door edge. When the door is pushed closed the ball is forced into a matching recess in the adjoining striker plate on the door frame, holding the door in place.

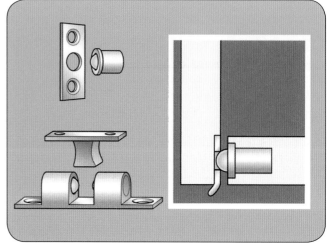

Door bolt

Door bolts are available in wide variety of styles and are fitted to one side of a twin door cabinet, with a cabinet lock fitted to the adjacent door.

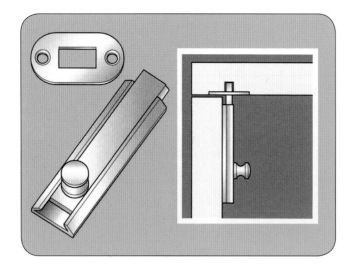

knock-down fittings

Knock-down (KD) fittings have revolutionized the furniture industry, paving the way for flat pack furniture – and no end of frustrating instruction leaflets. However, KD fittings also offer the wood-worker a great deal of flexibility and convenience when building projects. They are available for all types of jointing scenarios but really come into their own when a project will need to be disassembled and reassembled at some point in the future.

Chipboard insert

Chipboard is inherently weak when screwed into at the ends, but chipboard inserts function much like wall plugs because they expand to grip the chipboard when a screw is inserted and driven home.

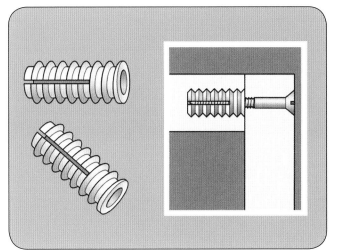

Related info

Project design (see pages 146–159)
Manufactured boards (see page 136)

Screw connector

Used to connect manufactured boards together, screw connectors feature coarse threads and are driven into pre-drilled holes with a hex key.

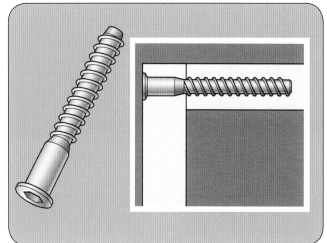

Using fittings

Knock-down fittings offer a great deal of versatility but tend to be associated with cheap furniture. However, there is no reason why you cannot use them on high quality furniture as well – and they can be especially useful when a project will need to be disassembled at some point in the future.

Cam fitting

Cam fittings are used to make corner joints in manufactured boards and consist of two components. A metal dowel, featuring a round Pozidriv head at one end and a screw thread at the other, is driven into one of two workpieces to be jointed. The second component consists of a boss with internal cam profile that is fitted into the second workpiece in line with the metal dowel. A connecting hole allows the dowel to be inserted into the cam boss, which is then rotated to grip the dowel and tighten the joint.

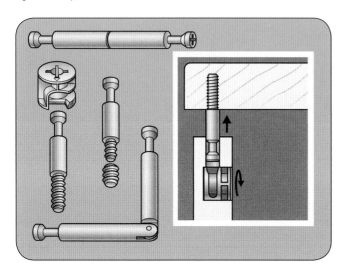

Threaded insert

Threaded inserts have a fine inside diameter thread – usually M6 or M8 – and a coarse exterior thread. A slotted or hex profile end allows them to be driven into a pre-drilled hole where they provide a strong fixing point for bolts and adjustable feet.

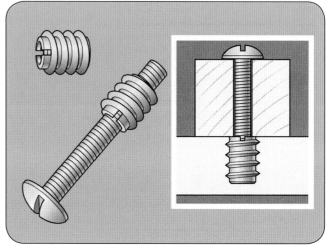

Pronged tee nut

Pronged tee nuts provide an alternative to threaded inserts but instead of being screwed into a hole they are driven in directly so that the prongs lock them in position. They are especially strong when joining components together.

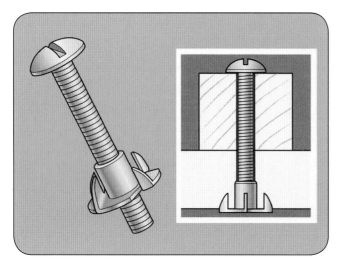

Panel connector

Used extensively to join sections of kitchen worktop together, panel connectors form a clamping joint between components and are visible from the underside only. Access and adjustment is possible via two access holes.

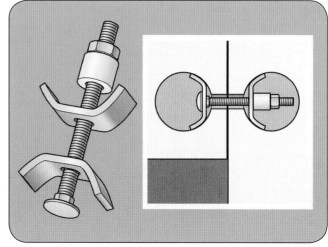

Cabinet connector

These connectors are mainly used to join kitchen cabinets and fitted bookcases together, and are very easy to fit.

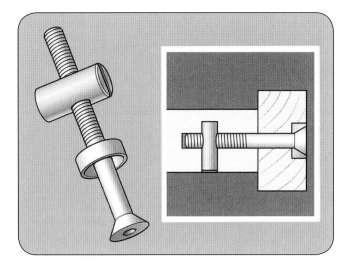

Bolt and barrel nut

Nicknamed bed bolts because of their extensive use in knock-down bed construction, bolt and barrel nuts provide a very strong clamping butt joint between end grain and long grain components. A hole is drilled through the long grain workpiece to take the bolt, and then continued into the second workpiece where an intersecting side hole is drilled to insert the barrel nut. The bolt is then driven through both workpieces and threaded through the barrel nut to tighten the joint.

Corner plate

Corner plates are used to join table legs to their corresponding rails, offering a fast knock-down alternative to traditional mortise and tenon joints.

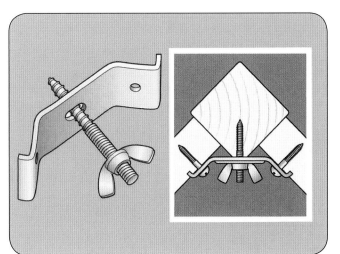

Shrinkage plates

These are widely used in place of traditional buttons for joining solid wood table tops to table under-frames. Shrinkage plates feature both slots and holes; the holes are for fixing to the stable under-frame, while the slots allow the table top to shrink and expand in width without splitting.

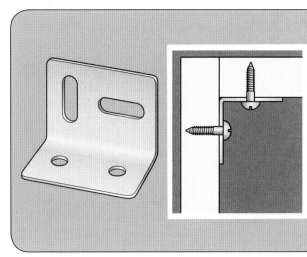

knobs and handles

Knobs and handles provide a gripping point to pull a door open or a drawer from its aperture and are available off-the-shelf or can be custom made. Knobs are usually turned round and fixed from the rear via a hole, whereas handles tend to be either recessed or lay flat on the surface, and can be fixed from either the front or the rear.

If making your own knobs, dowels can be used to invisibly fix them to a door or drawer front. Handles can be carved or drilled through surfaces if you would rather not fix a separate handle in place. Router cutters are also available for giving the lower edges of drawers a profile that is finger friendly and easy to pull.

Related info

Project design (see pages 146–159)
Hinges (see page 268)
Locks and catches (see page 271)

Choosing the right knob or handle

There are numerous types of knobs and handles available today. When choosing which is most appropriate for your project, consider the following questions:

1. Does my design look traditional or contemporary? Coming to a decision on this will quickly narrow down which type of fittings are best suited to your project.
2. Should I buy or make the handles? Often, making the handles yourself adds that extra-special ingredient to a project. Take this route if an overall handmade look is important to you.
3. Is quality important? There is a great deal of difference in quality between different types of knobs and handles. And price will be a major factor in your decision – the old adage "you get what you pay for" usually rings true here. Brass, for example, is more expensive than steel. However, steel is perfectly functional – it is just that brass tends to give a project a much more upmarket feel and does not corrode over time, so is more appropriate for good quality furniture.

screws and nails

Screws and nails provide a simple method of joining wood or manufactured board components and securing fittings in place quickly and effectively. Both are available in different forms optimized for combinations of materials. The primary difference between the two is that screws rely on a threaded shank for clamping force while nails rely on friction alone to hold components together.

Screws

There are many types of screw for different purposes – some are purely functional while others are also decorative. Screws are also available in different materials, such as solid brass, steel and stainless steel.

Slotted brass screws are often used in visible locations because of their attractive finish, although they can tend to shear easily. To avoid this, always drill pilot holes and apply a little candle wax to the thread before driving home. Alternatively, drive a steel screw of the same size home first, then remove it and replace with a brass screw – this minimizes the stress on the brass screw and so the risk of it shearing.

Steel screws are often hardened for strength and plated to prevent corrosion. They usually come with cross-head slots for easy and fast driving with a cordless drill/driver. Twin threaded screws, with their steeper thread pitch, increase the speed at which they can be driven while shankless variants – where the thread runs the entire length of the screw – provide an even stronger fix between components.

On acidic woods – such as oak – that react to steel fittings, stainless steel or brass screws should always be used.

How screws are measured

Along with head and thread type, screws have two defining measurements: length and diameter. Diameter is usually given as a gauge number from 0 to 20 (20 being the largest). Length is the measurement of the parts of the screw that are driven into the wood; so on a countersunk screw the entire length is included in the measurement because the head sits below the surface of the wood, while a round head screw is measured from the thread tip to the underside of the head since this will sit proud of the wood.

Choosing the correct screw length

Screw length is selected by first measuring the components to be joined. Ideally screw length should be three times the thickness of the piece to be secured, but if this is not possible a good compromise is to ensure that there is at least 3mm (⅛in) of material in front of the screw. This should be considered the minimum measurement to avoid a bump appearing on the opposite face as the screw compresses the wood fibres directly in front of it.

Related info

Screwdrivers (see page 59)
Hinges (see page 268)
Locks and catches (see page 271)

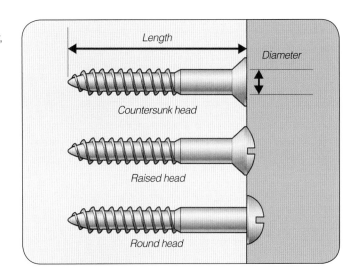

Length
Diameter
Countersunk head
Raised head
Round head

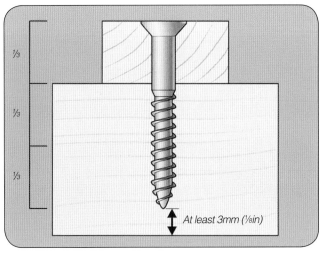

⅓
⅓
⅓
At least 3mm (⅛in)

Screw sizes

Length

Metric	Imperial	0	1	2	3	4	5	6	7	8	9	10	12	14	16	18	20
6mm	¼in	0	1	2													
9mm	⅜in		1	2	3			6		8							
12mm	½in			2	3	4	5	6	7	8							
16mm	⅝in				3	4	5	6	7	8		10					
18mm	¾in				3	4	5	6	7	8	9	10	12				
22mm	⅞in					4		6	7	8							
25mm	1in				3	4	5	6	7	8	9	10	12	14			
32mm	1¼in					4	5	6	7	8	9	10	12	14			
38mm	1½in					4		6	7	8	9	10	12	14	16		
44mm	1¾in							6	7	8	9	10	12	14	16		
50mm	2in							6	7	8	9	10	12	14	16	18	20
57mm	2¼in							6		8		10	12	14			
63mm	2½in							6		8	9	10	12	14	16		
70mm	2¾in									8		10	12	14			
75mm	3in							6		8		10	12	14	16	18	20
89mm	3½in									8		10	12	14	16		
100mm	4in									8		10	12	14	16	18	20
112mm	4½in											10	12	14			20
125mm	5in											10	12	14	16		
150mm	6in												12	14	16		

Screw heads vary in profile but usually fall into one of these three categories:

1. Countersunk; a countersink bit is used to allow the screw to sit flush with the wood surface.

2. Raised head; a countersink bit is used but the raised area of the head sits above the wood surface.

3. Round head; sits completely proud of the wood surface.

Nails

Although the use of screws, in preference to nails, has increased markedly over recent years, nails are still widely used for many woodworking and upholstery tasks. There are several different types available for an array of different applications.

1. Common round wire nails are used for general carpentry.
2. Oval wire nails, with their oval profile, are used to minimize splitting of the material.
3. Lost head nails are easily sunk beneath the surface of the wood with a punch, due to their diminutive head size.
4. Panel pins provide a neat solution for small nailing jobs.
5. Dedicated upholstery nails are designed to be visible and provide a decorative element to upholstery projects.
6. Gimp pins are used in upholstery to invisibly attach upholstery braid, usually to conceal edges or staple fixings.

Nail Categories

Type	Finish	Metric	Imperial	Use
Round wire nail	Bright steel	25–150mm	1–6in	Rough carpentry, mock-ups
Oval wire nail	Bright steel	25–150mm	1–6in	General purpose, oval shape reduces wood split, head punched below surface
Lost head nail	Bright steel	40–100mm	1½–4in	Finishing nail, large butt joints, mitres, head punched below surface
Panel pin	Bright steel	12–15mm	½–2in	Securing thin manufactured board, small joints
Corrugated fastener	Bright steel	6–22mm	¼–⅞in	Securing mitre and butt joints, driven flush with surface

Inserting screws

Do not confuse Pozidriv and Phillips pattern screwdrivers, which should be chosen carefully to fit the specific screw used – match the screwdriver to the screw head exactly.

Slippage can be a problem when inserting screws with a cordless drill/driver, although it is less of a problem with hand tools.

glossary

Ah (Amp hour) – battery capacity measurement. The higher the Ah rating, the longer the tool will run.

air drying – after sawing, boards are stacked in stick in the open air to dry naturally, covered against the rain. The rule of thumb is one year per 25mm (1in) of thickness. *See also kiln drying.*

armoured cable – electric cable for transferring mains power, with tough outer sheathing including wire reinforcement. Designed for outdoor use, it can be used overhead or buried underground at a set depth. The wire armour must always be earthed at one end.

bastard cut – file with coarse teeth (aggressive cut). *See also second cut and smooth cut.*

bed – the flat working surface or table of a machine. Also the lower structure of a lathe, either heavy twin steel bars or cast iron for rigidity.

bench hook – a rectangular hardwood base with upper and lower stops so it can be held against the bench edge. Used primarily to hold timber (lumber) steady when sawing.

between centres – maximum length of timber (lumber) that can be held between headstock and tailstock centres on a lathe.

biscuit – an oval shaped dowel of compressed beech, inserted into a matching slot in an adjacent board. *See also page 196.*

blank – a piece of rough–sawn timber (lumber), usually cut into a circle. Often used to turn a bowl when fitted to a lathe's faceplate.

blockboard – manufactured board material with a solid wood core (usually softwood) and veneered top and bottom faces.

book matching – two adjacent veneer leaves opened out and glued to groundwork to form a symmetrical pattern.

bpm – blows per minute, the measurement for impact drilling.

bridle joint – similar to a mortise and tenon joint, cut in reverse. *See also page 207.*

brush motor – electric motor commonly used in power tools and small machinery.

burnisher – used to form the burr on a cabinet scraper. Its hardened steel blade can be round or oval in cross section.

burr – growth on a tree trunk or substantial branch after damage has occurred. When cut off it reveals tight, wild grain and is usually sliced into veneer.

butt joint – simple joint where square ends are brought together and fixed – often pinned and glued. *See also page 184.*

caul – rigid board for cramping (clamping) groundwork and veneer together. May be curved or flat.

CITES – Convention on International Trade in Endangered Species.

chamfer – a bevel running along the edge of a piece of timber (lumber), normally at a 45–degree angle.

chuck – attached to the end of the spindle, with three or four self-centring jaws to grip a drill bit. On a hand drill or ratchet brace it is usually keyless and tightened by hand, though on older tools it may be tightened with key. *See also SDS.*

chuck – precision steel device for gripping a wooden blank instead of a faceplate.

clamp – US term for cramp.

collet – a split, tapered sleeve on a router to take the cutter shank. A locking nut secures it at the end of the motor spindle.

compound curves – a curve that runs in two planes simultaneously, for instance a splayed rear chair leg.

COSHH – Control of Substances Hazardous to Health Regulations.

crosscutting – sawing across the direction of the grain.

crown–cut – through–and–through timber (lumber) sawing, where growth rings seen in end grain are slightly curved. This produces crown-cut boards, With faces often displaying a flame pattern.

crown guard – sits above the exposed saw blade on a table saw to protect the user. Adjusts with the blade, can be made of plastic or metal.

cutterblock – a rotating cylindrical block with replaceable knives or cutters along its length. Made from steel or aluminium, a critical element of surface planers (jointers), spindle moulders and portable planers.

dado – a shallow groove running across the grain.

dead centre – similar to a revolving or live centre on a lathe but without ball bearings. Use grease or wax to lubricate.

de-nib – the process of sanding hardened areas of raised grain with fine abrasive paper.

double–cut – teeth criss-crossing over the face of a file. *See also single-cut.*

dovetail – a form of joint. *See also page 209.*

dowel bit – see lip–and–spur bit.

dowel joint – joint that uses accurately placed dowels as reinforcement. *See also page 197.*

drive centre – fitted in the headstock of a lathe, with two or four sharp prongs that are forced into end of spindle blank.

edge joint – form of butt joint where board edges are brought together and glued long grain to long grain – with or without reinforcement. *See also page 182.*

end grain – grain visible at the end of a piece of wood – at right angles to long grain. *See also long grain.*

epoxy resin – mixed with a hardener to form a very strong adhesive.

face edge – the second surface planed in timber (lumber) preparation, always at 90 degrees to the face side. Marking tools are subsequently held against either the face edge or face side.

face side – the surface planed first during timber (lumber) preparation. It should be completely straight and flat.

faceplate – steel disc fitted to a lathe for attaching wood blanks for bowl turning.

feed – the process of passing timber (lumber) or sheet materials through a machine, such as a saw.

feed speed – the speed at which timber (lumber) travels through a machine such as a thicknesser (thickness planer). The best finish is usually obtained at the slowest setting.

fettling – preparing a hand tool to perform more effectively. New budget tools often benefit from this process before sharpening.

figure – the grain pattern revealed in a piece of timber (lumber), usually when planed. Generally refers to unusual effects characteristic of certain woods.

flitch – log sliced into veneer.

former – shaped profile used to hold steam bent or laminated sections in place during the drying process.

French polish – traditional shellac–based finish used to achieve a high quality finish.

FSC – Forest Stewardship Council.

gallows bracket – a simple bracket, L–shaped with a bracing piece diagonally across the corner. Often made in the workshop.

grit – whether for hand or powered sanding, abrasive sheet is graded by particle size. The number refers to the amount of particles dropping through a mesh per unit area. The higher the number, the finer the grit.

groundwork – solid backing to which veneer is glued, often MDF or softwood.

guide bush – a steel collar screwed to the base of the router, enabling it to follow a template exactly. A cutter passes through the centre.

hardboard – cheap thin board material used for mock–ups and workshop rods.

hardwood – timber (lumber) from broad leaved trees, many of which are deciduous (shed leaves annually).

headstock – primary casting on a lathe, fitted at one end of the bed and housing spindle, belts and pulleys for adjusting speed.

heartwood – most stable timber (lumber) created by dense, hard cells at the tree's centre.

halving joint – two components halved in thickness then put together to form a joint of full thickness. *See also page 205.*

holdfast – a shaped metal rod that slides into a collar on the bench top. Tightened with screw adjuster or struck with hammer, it holds timber (lumber) in place rather like a cramp (clamp).

honing – the sharpening process for chisels and plane irons after grinding. A cutting edge is formed by honing the tool on a natural or synthetic stone.

housing joint (dado joint) – a joint in which a groove is cut in one workpiece to allow a second component to be recessed into it. *See also page 190.*

hp – horse power (1hp is the same as 760 watts).

HSS – high speed steel.

hygrometer – an instrument measuring relative humidity in a room, shown as a percentage.

in stick – boards stacked as a log is sawn, using stickers or battens at regular intervals.

induction motor – electric motor used in larger machines.

infeed – supporting table on a surface planer (jointer) or thicknesser (thickness planer), which the timber (lumber) passes first before reaching the knives. *See also outfeed.*

interlocking joints – joints featuring physical reinforcement such as biscuits or dowels.

jaws – on a lathe, the steel fingers of a chuck that grip a cylindrical wood blank by expanding or contracting. On a drill, the jaws grip the bit when it is inserted.

jig – a device used With a power tool or machine to cut or shape a timber (lumber) component safely and accurately. Often made in the workshop from wood or MDF.

KD (knock-down fitting) – off-the-shelf fittings for jointing and hinging components together in a way that is easily assembled and disassembled – used extensively in flat-pack furniture.

kerf – groove made in the wood by the teeth of a saw blade.

kickback – timber (lumber) thrown back from a machine towards the user by a spinning saw blade or cutterblock.

kiln drying – timber (lumber) seasoned in a kiln, usually by a process of dehumidification. Produces lower moisture content than air drying and is faster, used primarily for internal timber (lumber). *See also air drying.*

knives – removable blades or cutters on a planer. A portable power planer usually has two disposable knives, though some tools may have one spiral. Also used for matched blades or cutters on a surface planer (jointer) or thicknesser (thickness planer). On small machines there are normally two knives per cutterblock, with perhaps three or four on heavier machines.

laminating – sheets of veneer glued together to create a rigid shape, often curved. Plywood is made by laminating constructional veneers in sheet form.

lap joint – one component is rebated (rabbeted) to overlap another often pinned and glued. *See also page 187.*

Li-ion – Lithium-ion. See also NiCd and NiMH.

lint-free – free of loose fibres.

lip-and-spur bit – spiral twist drill with two outer spurs and sharp tip. Also known as a dowel bit.

long grain – grain running the length of wood at right angles to end grain. *See also end grain.*

manufactured boards – also referred to as man-made board. Wood-derived product manufactured in standard sheet sizes.

marking out – the process of marking waste areas and cut lines prior to cutting joints.

MDF – medium density fibreboard.

medullary rays – attractive flecks displayed when certain woods are quartersawn, particularly oak.**mechanical joints** – used to secure components in place while allowing controlled movement to occur – often necessary in solid wood construction.

mitre – an angled (usually 45-degree) joint between two pieces. *See also page 184.*

moisture content – weight of moisture shown as a percentage of the wood's dry weight.

morse taper – on a lathe or a pillar drill (drill press), the upper end of chuck terminates in a tapering spindle that fits tightly into the headstock.

mortise – a square or rectangular hole cut into wood, usually to take a tenon. A through mortise passes through the material completely, while a stopped mortise stops partway so the end of the tenon is concealed. *See also tenon.*

mortise and tenon – one of the most common woodworking joints. *See also page 200.*

MR – moisture resistant.

NiCd – Nickel cadmium. *See also Li-ion.*

NiMH – Nickel metal hydride

NVR – no volt release, the safety on/off switch fitted to any stationary machine.

opm – orbits per minute.

orbital – action of a sanding pad moving around a central point without spinning.

orthographic projection – a form of two dimensional drawing that incorporates three views in its standard form – front, top and side. Additional views can be created as required.

outfeed – support table on a surface planer (jointer) or thicknesser (thickness planer); timber (lumber) exits the machine on this table, having passed over or under the knives. *See also infeed.*

PAR – "planed all round", softwood planed square on every side.

paring – slicing away wood with a razor-sharp chisel. usually from the end grain.

pein – a wedge-shaped striking head on certain hammers, used to start driving small nails in more easily.

PIR – passive infra red device.

plane iron – more traditional term for the blade of a plane.

platen – cast alloy or plastic baseplate flexibly mounted to an orbital sander.

plywood – manufactured board made from layers of veneer laminated together with opposing grain directions.

pocket hole joint – method of concealed screwing, often created with the use of a dedicated jig. *See also page 198.*

ppi – points per inch (25mm).

pushstick – a wood or plastic safety device for pushing narrow or small components past a table saw blade. It prevents fingers getting too close to the moving blade.

PVA (Polyvinyl acetate) – water based adhesive used extensively in woodworking.

quartersawn – boards sawn with growth rings at least 45-degrees to the surface, which produces stable timber (lumber) with minimal shrinkage.

rabbet – US term for a rebate.

RCD – residual current device.

rebate – square section recess cut into the edge of a workpiece. The US term is rabbet.

revolving or live centre – fitted in the tailstock of a lathe, the point is inserted into end of a wood blank mounted on a spindle. Ball bearings enable it to revolve smoothly.

ripping (rip sawing) – cutting parallel to (or with) the grain of the timber (lumber).

riving knife-shaped steel arm behind saw blade, wider than the bladethickness but thinner than the kerf, to prevent kickback by stopping wood closing up and binding on the rotating blade.

rod – see workshop rod.

rough-sawn – wood is rough-sawn during the conversion process prior to being planed and thicknessed.

rpm – revolutions per minute.

rub joint – butt joint made by rubbing two flat component edges to create a vacuum while the glue sets. *See also page 182.*

rubber – used in French polishing to apply polish.

sacrificial board – off-cut of plywood, MDF or timber (lumber) used as a renewable surface to support a workpiece when drilling or sawing. Ensures a clean cut and prevents bit or blade contacting metalwork of machine.

sapwood – new wood, the least stable timber (lumber) created by soft cells growing furthest from the tree's centre.

scarf joint – a method of jointing wood end to end to increase its length. *See also page 189.*

screwdriver bit – small bit with ¼in hex shank used with cordless drill drivers. Normally fitted into a spring-loaded bit holder for fast changeover. Wide range of screw patterns.

SDS (Special Direct System) – clamping system where bits lock automatically in the chuck without having to tighten them

seasoning – process where moisture is removed from sawn timber (lumber) cell walls, either by air or kiln drying.

second cut – file with medium teeth. *See also bastard cut and smooth cut.*

set – saw teeth are alternately bent slightly to one side, then the other. The resulting cut (kerf) is wider than the blade itself.

setting out – the process of organizing wood, marking it and cutting it to approximate length prior to planing it square.

shadow board – panel fixed to the wall with hooks for storing hand tools. Often painted white with tool outlines in another colour.

shellac – animal substance derived from an insect, *Laccifer lacca,* in India and used in the manufacture of French polish.

shiplap – often used as horizontal cladding on wooden buildings. Made of overlapping softwood boards – the rebate (rabbet) on the bottom edge sits against top edge of the adjacent board.

shoe – heads on a cramp (clamp) that exert pressure on the workpiece. Usually fixed at one end, while the other is adjustable.

shooting board – workshop-made; used at a bench to guide a bench plane when trimming narrow pieces and squaring end grain.

single-cut – teeth sloping in one direction across the face of a file. *See also double-cut.*

smooth cut – file with finest teeth. *See also bastard cut and second cut.*

softwood – timber (lumber) from coniferous trees.

sole – ground lower face of a plane or spokeshave that moves across the timber (lumber). A spokeshave sole is convex, concave or flat.

spindle – a revolving shaft passing through the headstock of a lathe, fitted in bearings.

spindle moulder – a stationary machine fitted with shaped knives for cutting profiles, edges and joints in straight or curved timber (lumber). Highly accurate, but can be dangerous without training.

spm – strokes per minute.

stock – hardwood piece that accepts the sliding stem (on a gauge) The piece that anchors the blade (on a square). Usually hardwood, sometimes metal or plastic.

stock – timber (lumber) planed square all round ready to use.

stopped joint – a joint cut part of the way through the thickness of the component so it so it is not visible or exposed on the outer face, for example a stopped housing. *See also through joint.*

stroke length – the distance a jigsaw or reciprocating saw blade moves up and down.

swing – the largest diameter wood blank that fits on a lathe without contacting the bed.

tailstock – cast iron support at the opposing end to the headstock of the lathe. Used if turning between centres, locked to support spindle. Moves along the bed when released.

tang – the tapered end of a rasp or file, driven into a handle.

taper-ground – on top quality handsaws the blade is ground thinner towards the top to reduce friction when cutting.

TCT – tungsten carbide tipped.

temper – to harden a tool's cutting edge by heating, then cooling the steel.

tenon – male part of a mortise and tenon joint. *See also mortise.*

thicknessing – process of reducing a section of wood to a desired and uniform thickness, usually by machine.

throat capacity – the distance from the centre of the adjuster shoe to the neck or bar of a G or F cramp (clamp). Also the maximum cutting width of a bandsaw.

through joint – a joint that is cut all the way through a component with its exit point visible. *See also stopped joint.*

tongue and groove – planed softwood boards that lock together. One edge has a tongue, the other has a corresponding groove. Also called match boarding.

tool rest – horizontal cast iron bar against which tools are held when turning. Adjusted for height and angle, it slides along the lathe bed and is locked in position.

torque – the twisting force when inserting a screw. Measured in Nm (Newton meter) in power tools.

tpi – teeth per inch (25mm).

tracking – Sideways movement of the abrasive belt across a belt sander's rollers. Also the path a narrow blade takes around the wheels of a bandsaw.

trenching – a groove sawn across the grain, rather than parallel to it. Also known as a housing, its depth is regulated by setting saw blade height and locking in position.

trunnion – on a bandsaw frame the table is bolted to a cast trunnion. A protractor scale and locking levers enable the table to be tilted at set angles.

veneer – thinly sliced wood, which is often stuck to a cheaper substrate to resemble thicker wood. Often the only economical way to use exotic woods.

waney edge – The natural outer edge of a board, often with bark still attached.

winding sticks – twin matching hardwood or metal strips used to check if a surface is flat or twisted (in wind). Both should be dead straight and have parallel edges.

workshop rod – full size 2D drawings usually created on hardboard to provide a highly accurate 1:1 scale reference that can be used to take measurements from at the bench.

WWF – World Wide Fund for Nature.

suppliers and websites

TOOLS

Axminster Tool Centre
Unit 10 Weycroft Avenue, Axminster,
Devon EX13 5PH
Tel: 0800 371822
www.axminster.co.uk
Hand tools, power tools, machinery.

BriMarc Tools & Machinery
Unit 10 Weycroft Avenue, Axminster,
Devon EX13 5PH
Tel: 0333 2406967
www.brimarc.com
Hand tools and machinery.

Classic Hand Tools
Hill Farm Business Park
Witnesham, Ipswich
Suffolk IP6 98N
Tel: 01473 784983
www.classichandtools.co.uk
Hand tools, books, DVDs.

Richard Maguire Traditional Workbenches
Tel: 07920 423123
www.rm-workbenches.co.uk
Superb hand-made workbenches.

Pennyfarthing Tools
Tel: 01235 763987
www.pennyfarthingtools.co.uk
Old woodworking tools.

Philly Planes
www.phillyplanes.co.uk
Fine hand-made wooden planes.

Tilgear
Langley House,
Station Road
Standon, Herts SG11 1QN
Tel: 0845 099 0220
www.tilgear.org
Hand and power tools.

Workshop Heaven
Tel: 01295 678941
www.workshopheaven.com
Hand tools.

TIMBER (LUMBER), VENEER, BOARDS

Art Veneers/Craft Supplies
Tel: 01302 744344
www.craft-supplies.co.uk
Veneers, tools, woodturning.

Avon Plywood
Pixash Works, Pixash Lane,
Keynsham, Bristol BS31 1TR
Tel: 0117 986 1383
www.avonplywood.co.uk
Manufactured boards.

Capital Crispin Veneer
12 & 13 Gemini Business Park,
Hornet Way, Beckton,
London E6 7FF
Tel: 020 7474 3680
www.capitalcrispin.com
Veneers.

Morgan Timber
Knight Road, Rochester,
Kent ME2 2BA
Tel: 01634 290909
www.morgantimber.co.uk
Hardwoods, softwoods, boards.

Timberline
Unit 7, Munday Industrial Estate,
58–66 Morley Road, Tonbridge,
Kent TN9 1RP
Tel: 01732 355626
www.exotichardwoods.co.uk
Exotic hardwoods and softwoods, veneers.

Yandles
Hurst Works, Hurst
Martock, Somerset TA12 6JU
Tel: 01935 822207
www.yandles.co.uk
Hardwoods and softwoods, veneers, marquetry, hand tools, power tools, machinery, woodturning equipment.

FINISHING PRODUCTS

Behlen
Tel: 0115 973 7288
www.behlen.co.uk

Chestnut Products
Tel: 01473 890118
www.chestnutproducts.co.uk

Fiddes
Tel : 029 2034 0323
www.fiddes.co.uk

Liberon
Tel: 01797 367555
www.liberon.co.uk

Rustins
Tel: 0208 450 4666
www.rustins.co.uk

HARDWARE & FITTINGS

Ironmongery Direct
Tel: 03332 228 199
www.ironmongerydirect.com
Fittings, hardware.

Screwfix
Tel: 03330 112 112
www.screwfix.com
Hardware, hand and power tools.

MAGAZINES

Good Woodworking
The Woodworker
Practical Woodworking
www.getwoodworking.com
Britain's leading woodwork magazines.

index

acknowledgements

The authors would like to thank the following companies and individuals for their help with this book. Without them there would have been many blank spaces!

For images of HAND TOOLS:
Axminster Tool Centre – www.axminster.co.uk
Bahco – www.bahco.com
BriMarc – www.brimarc.com
Stanley – www.stanleyworks.co.uk
Phil Edwards – www.phillyplanes.co.uk

For images of POWER TOOLS:
AEG Power Tools – www.aeg-pt.com
Axminster Tool Centre – www.axminster.co.uk
Black & Decker – www.blackanddecker.eu
Robert Bosch – www.boschpowertools.co.uk
DeWalt – www.dewalt.co.uk
Dremel – www.dremeleurope.com
Makita – www.makitauk.com
Metabo – www.metabo.co.uk
Milwaukee – www.milwaukeetool.com
Ryobi – www.ryobipower.co.uk
Trend – www.trend-uk.com
Triton – www.tritontools.com

For images of MACHINES:
Axminster Tool Centre – www.axminster.co.uk
BriMarc – www.brimarc.com
Record Power – www.recordpower.co.uk
Startrite – www. startrite.co.uk

For images of MATERIALS:
Morgan Timber – www.morgantimber.co.uk

Thanks to Robert Sorby for loaning turning tools and woodturner Dave Roberts for his beautiful turned boxes. Also thanks to Atkinson Walker for providing saw blades.
www.robertsorby.co.uk
www.atkinson-walker-saws.co.uk

Many thanks to Art Veneers/Craft Supplies and Capital Crispin Veneer for supplying veneers and to Avon Plywood for board samples.
www.craft-supplies.co.uk
www. capitalcrispin.com
www. avonplywood.co.uk

Jacket photograph and style photography by Ben Plewes, www.benplewes.com
Photographs on pages 14, 22, 28 top left, 30 bottom right, 31, 38 right, 41 bottom right, 42 centre, 53, 74 top right, 82 top right, 83 right, 99 bottom right, 110 bottom right, 123, 125, 126, 127, 128, 130, 132 and 133 by Phil Davy.

Special thanks to Marie Clayton – without her this book would probably never have been completed! Thanks also to Kuo Kang Chen for his excellent illustrations, Louise Leffler for her great design, and everyone involved at Collins & Brown.

Finally, a few words from Ben: "Thanks to my wife Steph for her continual support and to our daughter Freya for reminding us of what's important in life ... "

And from Phil: "Cathy, thanks for your encouragement and patience! And grateful thanks to the Master Craftsman ... "

Phil Davy studied musical instrument making at the London College of Furniture and has taught carpentry and joinery at colleges in the west of England. As a qualified wood machinist, Phil is a Consultant Editor for *Good Woodworking* magazine.

Ben Plewes was trained as a furniture maker in the early 1990s before working for a number of London-based design consultancies and furniture manufacturers. Ben currently runs his own bespoke furniture making business.

Also in this series:

978-1-910231-76-0

978-1-910231-77-7

978-1-910231-79-1

978-1-90844-901-6

978-1-90939-718-7

978-1-911163-42-8

The craft of woodworking has been practised for centuries. Being able to create a well-made wooden item is a useful skill and can also be intensely satisfying. *Ultimate Woodwork Bible* is a fully illustrated and comprehensive guide, which not only covers all the basic woodworking techniques in detail but is also crammed with other useful information to help you learn how to transform a simple piece of wood into a beautiful or ornamental object.

The first chapter covers the process of choosing a suitable space in which to work, how to plan a workshop, and essential safety issues – as well as tips on workshop security. Following this are chapters on hand tools, power tools, and woodwork machines, with details of the different types available and factors to consider before purchasing. The section on materials looks at the characteristics of different types of wood, and explains what to be aware of when buying timber (lumber) or manufactured boards.

Moving on to woodworking skills, there is a chapter on designing and planning projects – right from getting inspiration to preparing a cutting list – and then sections covering all the different techniques, including wood preparation, construction, adhesives, bending and shaping, finishing, and choosing hardware. All the fundamental skills are illustrated with clear step-by-step instructions and diagrams, which cover both the techniques for hand tools and for working with machinery, so you have the option of choosing the method that best fits your tooling and skill level.

As well as specific instructions, *Ultimate Woodwork Bible* is also full of useful professional tips and safety information for easier and safer woodwork. Whether you are a novice or an experienced woodworker, it will become an indispensable reference source that you will use time and time again, either to learn a new technique or to remind you of a specific detail.